Computer Communications and Networks

The **Computer Communications and Networks** series is a range of textbooks, monographs and handbooks. It sets out to provide students, researchers and non-specialists alike with a sure grounding in current knowledge, together with comprehensible access to the latest developments in computer communications and networking.

Emphasis is placed on clear and explanatory styles that support a tutorial approach, so that even the most complex of topics is presented in a lucid and intelligible manner.

Also in this series:

Ana-Belén García-Hernando · José-Fernán
Martínez-Ortega · Juan-Manuel
López-Navarro · Aggeliki Prayati ·
Luis Redondo-López, MsC

Problem Solving for
Wireless Sensor Networks

 Springer

Ana-Belén García-Hernando, PhD
Universidad Politécnica de Madrid
Spain
abgarcia@diatel.upm.es
anabelen.garcia@upm.es

José-Fernán Martínez-Ortega, PhD
Universidad Politécnica de Madrid
Spain
jfmartin@diatel.upm.es
jf.martinez@upm.es

Juan-Manuel López-Navarro, PhD
Universidad Politécnica de Madrid
Spain
jmlopez@sec.upm.es
juanmanuel.lopez@upm.es

Aggeliki Prayati, PhD
ATHENA / I.S.I. Research and
   Innovation Center in Information
Communication, and Knowledge
   Technologies, Greece
prayati@isi.gr

Luis Redondo-López, MsC
Métodos y Tecnología de Sistemas y
   Procesos (MTP), Spain
lredondo@mtp.es

ISBN: 978-1-84800-202-9   e-ISBN: 978-1-84800-203-6
DOI 10.1007/978-1-84800-203-6

British Library Cataloguing in Publication Data
A catalogue record for this book is available from the British Library

Library of Congress Control Number: 2008926114

Printed on acid-free paper

Springer Science+Business Media
springer.com

Preface

Book's Overview and Features

Wireless sensor networks (WSNs) have quickly become an area of great interest in terms of research for both industry and academia. Nowadays, the enormous potential of this technology can be easily seen, along with its inherent difficulties. Just looking at the number of research projects being funded, mainly European- and U.S.-based, the many research papers being published, and the results being put on the market gives clear evidence of the technology's growing importance. In fact, the Massachusetts Institute of Technology recently classified WSNs as one of the 10 emerging technologies that will change the world.

This book is the result of intensive research carried out over several months as part of a European research project. It constitutes a wide review of the current state of the art regarding wireless sensor networks at the time of its writing. Contributions have been made by several researchers from various organizations.

Other research teams and European projects have also made very valuable contributions in the field of wireless sensor networks. However, to the best of our knowledge, this book is the only one encompassing all of the following characteristics:

- It is entirely dedicated to wireless sensor networks and comprises all of the main technological challenges associated with them: from hardware to specific applications, including networking, middleware, and software issues.
- It not only includes a review of commercially available products and solutions, but also examines European research projects concerning WSNs and open issues currently of interest for researchers in this area. Moreover, there is also a chapter devoted to regulatory and safety issues related to this technology.
- It includes a description of several exemplifying application scenarios in which the use of a WSN solution is very attractive, something that may inspire current and future applications.

Target Audiences

All of these features make the book useful for a wide range of potential readers, including researchers in the computer/wireless communications sector, lecturers for advanced communication courses, graduate students beginning research in computer/wireless communications, professionals wanting to offer WSN solutions, and even WSN application designers.

Our aims are to help the reader grasp the main technological issues to be considered when dealing with WSNs, to give a high-level overview of the different technologies available, and to pave the way for an eventual deeper study of specific aspects of this wireless technology.

Acknowledgments

This book has been written mostly as part of the work done in the European research project "Solving Major Problems in MicroSensorial Wireless Networks" (μSWN), of the Sixth Framework Programme of the European Union. We would like to thank the people of the μSWN Consortium who have contributed to the quality of this book either by writing specific parts of it or by reading and giving feedback. The figures and technical information related to Wavenis technology and products present in the book have been provided by Coronis Systems and are reproduced with permission.

We would also like to thank the project officer in charge of the μSWN project (Rolf Riemenschneider) and the project technical reviewers (Prof. Luis Orozco Barbosa, Universidad de Castilla La Mancha, Spain; Dr. Christoph Niedermeier, Siemens AG, Germany; and Dr. Gilles Thonet, Schneider Electric SA, France) for encouraging us to publish this work in order to reach a wider audience.

We acknowledge the support we have obtained from Springer-Verlag London Ltd. to publish the book, particularly Mr. Wayne Wheeler (Senior Editor, Computing & Information Science), who received our proposal and gave us prompt feedback, and Ms. Catherine Brett (Senior Editorial Assistant) who always answered our queries with precision and has been very supportive during the editing process. Ms. Tenmogi Sinnaveerappan, project manager at Integra Software Services, has been in charge of the final production process of the text, and we would like to thank her for her helpful guidance during this last phase.

Editors' and Contributors' Note

Since this book contains mainly a state-of-the-art report on a research field, a wide variety of information from very diverse sources has been read, compiled, and presented. This information includes research papers, books on different subjects, Web pages of varying nature, and manufacturers' products brochures, among other material.

To the best of our knowledge, this information is accurate with respect to what the corresponding authors claim. However, we cannot guarantee its correctness nor accept any responsibility for any damage of any type derived from its use, since we are not the primary source of the information and misunderstandings or errors may have occurred. We encourage those interested in obtaining guaranteed information to contact the sources mentioned directly, especially when dealing with product manufacturers.

Contents

Contributors

Christos Antonopoulos
Industrial Systems Institute, Greece

Fabrice Auzanneau
CEA LIST, France

Yannick Bonhomme
CEA LIST, France

Mickael Cartron
CEA LIST, France

Iván Corredor-Pérez
Universidad Politécnica de Madrid, Spain

Ausra Dagilyte
Birstonas Municipality, Lithuania

Antonio da Silva-Fariña
Universidad Politécnica de Madrid, Spain

Guillermo de Arcas-Castro
Universidad Politécnica de Madrid, Spain

Ana-Belén García-Hernando
Universidad Politécnica de Madrid, Spain

Valentin Gherman
CEA LIST, France

Spilios Giannoulis
Industrial Systems Institute, Greece

Vicente Hernández-Díaz
Universidad Politécnica de Madrid, Spain

Eduardo Hernández-Pérez
Universidad de las Palmas de Gran Canaria, Spain

Fotis Kerasiotis
Industrial Systems Institute, Greece

Christos Koulamas
Industrial Systems Institute, Greece

Agnius Liutkevicius
Kaunas University of Technology, Lithuania

Álvaro Llorente-Alonso
Universidad Politécnica de Madrid, Spain

Javier Longares-Abaiz
Edosoft Factory, Spain

Mario López-Marcos
Universidad Politécnica de Madrid, Spain

Juan-Manuel López-Navarro
Universidad Politécnica de Madrid, Spain

Lourdes López-Santidrián
Universidad Politécnica de Madrid, Spain

Fernando D. Lorenzo-García
Universidad de las Palmas de Gran Canaria, Spain

Emily Louisa Manning
Métodos y Tecnología, Spain

José-Fernán Martínez-Ortega
Universidad Politécnica de Madrid, Spain

Christophe Maugenest
CORONIS, France

Juan L. Navarro-Mesa
Universidad de las Palmas de Gran Canaria, Spain

Aggeliki Prayati
Industrial Systems Institute, Greece

Daniel Quijano-Díaz
Edosoft Factory, Spain

Pedro J. Quintana-Morales
Universidad de las Palmas de Gran Canaria, Spain

Manuel Ramiro-Mauleón
Métodos y Tecnología, Spain

Miguel Ramos-Herrero
Métodos y Tecnología, Spain

Luis Redondo-López
Métodos y Tecnología, Spain

Luis-Daniel Rosado-Poveda
Universidad Politécnica de Madrid, Spain

Mariano Ruiz-González
Universidad Politécnica de Madrid, Spain

Tsenka Stoyanova
Industrial Systems Institute, Greece

Juan-Alberto Vera-Gómez
Edosoft Factory, Spain

Arunas Vrubliauskas
Kaunas University of Technology, Lithuania

Chapter 1
Introduction

1.1 Executive Summary

The main goal of this book is to describe and analyze the technological novelties of wireless sensor networks (WSNs) from software and hardware points of view. We highlight the main difficulties when using WSN technologies and identify unsolved problems, current open issues in research, and challenges in this area. This includes hardware solutions, hardware platforms and radio technologies, network deployment aspects, communication protocols, middleware solutions, and interconnection of WSNs with other networks. In addition, we describe the characteristics and requirements of three representative scenarios to demonstrate the advantages of the wireless sensor technology and its boundless potential in today's world. Regulatory and safety issues related to WSN have their own dedicated chapter.

Part of the contents of this book is the result of work done inside the μSWN European research project ("Solving Major Problems in Microsensorial Wireless Networks," FP6-2005-IST) (uSWN), in which the editors and contributors of this volume have participated.

We are aware that other research teams and European projects have already made very valuable state-of-the-art contributions to wireless sensor networks. To mention two of the most important and recent works, we can cite the documents of the "Embedded WiSeNts" research project, and particularly the "Embedded WiSeNts Research Roadmap" report (Marrón et al., 2006), and the "Sensor Networks Roadmap & Strategic Research Agenda" of the CRUISE project (Ashraf et al., 2006). Although we have considered these projects, the information contained in our book is aimed at fulfilling goals directly applicable to solving problems in WSN. We have not tried to produce yet another medium- to long-term, explicit roadmap of WSN research activities. Rather, we highlight the major open issues that still have to be tackled to make WSN reach their full foreseeable great potential.

A.-B. García-Hernando et al., *Problem Solving for Wireless Sensor Networks*,
DOI: 10.1007/978-1-84800-203-6_1, © Springer-Verlag London Limited 2008

1.2 Structure of the Book

Seven chapters follow this introduction. The first four chapters are devoted to analyzing the state of the art regarding radio-frequency technologies, hardware platforms, software technologies, and network aspects of wireless sensor networks, respectively. Chapter 6 discusses the main safety and regulatory issues related to WSNs, while Chapter 7 gives an overview of the main recent and ongoing European research projects related to this technological area. The final chapter describes some relevant scenarios in WSNs, including the user requirements and a technical description of the network infrastructure and the overall system that should be designed to support them. Each chapter finishes with the relevant bibliographic sources to be checked if more precise information is sought.

Chapters are self-contained, thus making it possible for readers to focus on their specific interests without the need to read the chapters in any predefined order. Especially significant are the "open issues" sections located in most chapters. These sections contain what we have identified as major research challenges that still exist in the WSN area, after reviewing many scientific and technical documents and papers on the subject of each chapter. This information was very useful for partners in our European project to concentrate our research efforts on solving actual current "problems" in wireless sensor networks. We would really like them to inspire future research initiatives in this exciting field.

References

Ashraf I, Todorova P, Aguero R, et al. (2006) Sensor Networks Roadmap & Strategic Research Agenda. CRUISE project deliverable D113.1.
Marrón PJ, Minder D, the Embedded WiSeNts Consortium (2006) *Embedded WiSeNts Research Roadmap*. Logos Verlag, Berlin.
uSWN Project Consortium. Solving major problems in microsensorial wireless networks. https://www.uswn.eu.

Chapter 2
Radio-Frequency Technologies for WSNs

Abstract Many wireless monitoring and control applications are available for the industrial and home markets. Some hardware platforms are specialized for optimizing only one feature (e.g., high data rate, long transfer range, or low-power mode). However, the most restrictive parameters for WSNs are both power consumption and distance. This chapter briefly describes different radio-frequency technologies, although many of them are not appropriate or are not yet fully developed for WSNs. Therefore, only appropriate technologies are discussed in depth, and a brief overview of several integrated circuits from different manufacturers is included.

2.1 Bluetooth Technology (IEEE 802.15.1)

The Bluetooth wireless communications technology provides a personal area network (PAN) for exchanging data between Bluetooth-capable devices within a certain proximity.

Bluetooth technology has a low-power mode and high integrated devices and operates in the unlicensed 2.4-GHz band, but it is limited to short-distance communications. Therefore, this technology is not the most appropriate for developing a WSN. For this reason, Bluetooth is just mentioned and described as an existing technology (Bluetooth SIG).

2.2 Wi-Fi Technology (IEEE 802.11.a/b/h/g)

The Wi-Fi technology allows different devices like laptops, personal computers (PCs), cell phones, and personal digital assistants (PDAs) to communicate between one another or to connect to the Internet without needing a cable connection.

The Institute of Electrical and Electronics Engineers (IEEE) defined the Wi-Fi network protocols IEEE 802.11a, 802.11b, and 802.11 g operating in the unlicensed radio bands of 2.4 and 5 GHz. Therefore, any kind of standard

A.-B. Garcia-Hernando et al., *Problem Solving for Wireless Sensor Networks*,
DOI: 10.1007/978-1-84800-203-6_2, © Springer-Verlag London Limited 2008

Wi-Fi–certified device is able to operate all over the world with data rates of 11 Mbps for IEEE 802.11b or 54 Mbps for IEEE 802.11a. Of course, the greater the distance to the access point, the lower the performance.

Wi-Fi technology lacks a low-power mode and is also not very highly integrated. Thus, a low-powered and highly integrated WSN cannot use this technology, which is why we give just an overview of this existing technology (Wi-Fi Home Page).

2.3 UWB Technology (IEEE 802.15.3)

The Ultra Wideband (UWB) technology allows information to be transmitted at a large bandwidth in precise pulses that are typically 1 to 2 nanoseconds in length and occupy at least 25% of the center frequency, much more than other systems. The use of this technology is limited to the range of frequencies from 3.1 to 10.6 GHz. Another remarkable characteristic of UWB is its better behavior regarding interferences than other technologies due to the use of spread-spectrum modulation techniques.

Despite having a greater ratio of transmission velocity over power consumption than other similar technologies like Wi-Fi, UWB is limited to short-range applications. It is therefore appropriate for portable devices, to get long battery life, but not for WSNs requiring larger distances. This is the main reason for not going into further details on UWB technology (Intel UWB).

2.4 Wavenis Technology (EN300–220 and FCC15.247—Coronis Systems)

The Wavenis technology, developed by Coronis Systems, provides long-range data connections and services for autonomous devices with extremely limited battery resources and is intended for ultra low-power (ULP) and long-range wireless communications. Wavenis extends the industry standard Bluetooth protocol to provide robust wireless solutions for building ad hoc and fixed networks using autonomous, battery-powered devices.

2.4.1 Wavenis' Main Characteristics

Wavenis is a complete software and hardware solution for wireless communications in low-power devices. The core offer consists of an RF transceiver and a protocol stack, both specifically adapted to provide the optimal combination of secure and reliable connections, long range, and minimal power consumption. Highly resistant to interference and obstacles, Wavenis offers a means for including a wide variety of battery-powered products in PAN, LAN, and WAN networks, particularly for low-data-rate domestic and industrial applications. Wavenis implements an advanced RF architecture and can serve as a means to extend Bluetooth applications toward applications for which no other solutions address real-world needs so directly. This extension not only ensures interoperability with existing networks, but also complements other protocols and avoids costly proprietary gateway solutions. The main features of Wavenis are

- Small footprint
- Low power consumption
- Low per-unit cost
- Line-of-sight connections up to 1 kilometer
- Completely programmable configurations
- Easy in-field installation
- Coexistence with other technologies; interference- and obstacle-resistant
- Point-to-point, broadcast, polling, repeater functions, mesh network topologies
- Ongoing QoS monitoring to have reliable communications

2.4.2 Wavenis' RF ASIC Solution

Wavenis licensees may benefit from the Wavenis ASIC, specific low-power baseband, and a full control over system design and RF integration to best meet the designer product goals. This offers the following features:

- It is a flexible solution for mass-produced products.
- Designers can build their own radio interface based on the Wavenis RF ASIC.
- Designers can run the Wavenis protocol stack in their own microcontroller.
- It has a low per-unit license cost.
- It has a small hardware and software footprint, for a low Bill of materials.
- It is stack-portable to many microcontrollers.
- The stack object code is available.

Figure 2.1 shows the structure of the Wavenis layers. The HCI interface offers a simple API to use Wavenis services. The designers can find services such as object code and APIs, Active X controls for Windows applications, and Wavenis' stack configuration tools.

Wavenis operates at frequency bands of 868, 915, and 433 MHz, with data rates between 2.4 and 100 kbps (typical 10 kbps), and uses different techniques for saving power and making the network robust, such as data interleaving, FHSS (frequency-hopping spread spectrum), forward error correction, and GFSK modulation. Wavenis has a high line-of-sight range of up to 1 kilometer and automatic frequency control (AFC) to guarantee high performance levels over the product's operating lifetime. Also, it is compliant with relevant European and U.S. electromagnetic compatibility regulations.

Fig. 2.1 Wavenis' protocol stack

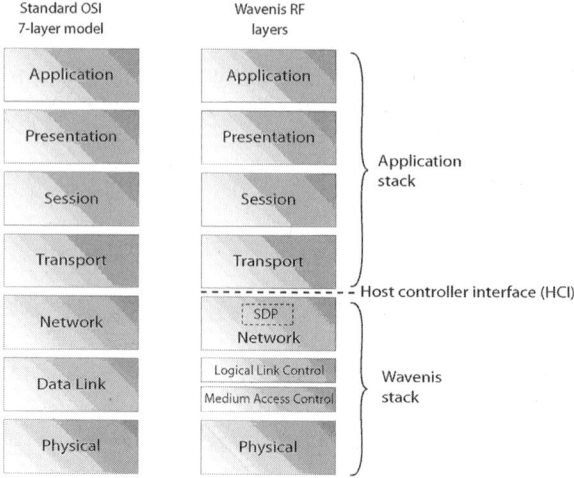

Coronis is also currently developing its new system-on-chip ASIC. The new solution aims to offer unprecedented low-power consumption for large-scale wireless mesh networks and higher RF performance.

2.5 Wibree Technology (Nokia)

Wibree technology is a short-range wireless communications protocol intended to compliment Bluetooth by implementing most of the Bluetooth functions with less power consumption. The Wibree open standard is able to work in applications where reliable Bluetooth data transmission is not possible, although the maximum data transmission rate is three times lower than that of Bluetooth 2.0 (1 Mbps versus up to 3 Mbps).

The Wibree specifications are being defined by a group of important companies from different sectors such as semiconductor manufacturers, service providers, and vendors, with Nokia in the lead. (The list of such companies includes Broadcom, Casio, CSR, Epson, ItoM, Nordic Semiconductor, STMicroelectronics, Suunto, Taiyo Yuden, and Texas Instruments.)

This technology is designed to operate with either a standalone chip or a dual-mode chip. While the standalone chip is a small device able to operate with very low power consumption, the dual-mode Bluetooth Wibree is able to communicate with Bluetooth standard devices with less power consumption and at distances of 5 to 10 meters using the 2.5-GHz band.

The main characteristics of Wibree are the ultra low-power IDLE mode operation, power-saving technology, device discovery, reliable point-to-multipoint data transfer, and encrypted communications (Nokia, 2007).

Wibree is not explained here in more detail because it is not a long-distance technology, therefore being suited for low–power, small devices at limited distances (5–10 meters), not totally adequate for many typical WSN applications (Wibree Home Page).

2.6 ZigBee Technology

The ZigBee technology is a communications standard for systems with requirements such as long battery life, low data rates, secure communications, and less complexity compared with previous wireless standards. It is based on the IEEE 802.15.4 standard (IEEE, 2006) for wireless personal area networks (WPANs).

2.6.1 ZigBee's Main Characteristics

IEEE 802.15.4 is a protocol for wireless networks aiming to achieve simplicity, low cost, low data rate, and low power consumption with the ability to operate months or even years with standard AA or AAA batteries (Freescale ZigBee Overview).

The IEEE 802.15.4 standard defines two layers, the MAC and the physical layer (PHY), as shown in Fig. 2.2, and uses the three license-free frequency bands. These license-free bands have a total of 27 channels divided into 16 channels at 2.4 GHz with data rates of 250 kbps, 10 channels at 902 to 928 MHz with data rates of 40 kbps, and one channel at 868 to 870 MHz with a data rate of 20 kbps. However, only the 2.4-GHz band operates worldwide; the others are regional bands. The 868–870-MHz band operates in Europe, while the 902–928–MHz band operates in North America, Australia, and other countries (ZigBee Home Page).

IEEE 802.15.4 has adopted the direct-sequence spread-spectrum (DSSS) technique in order to ensure coexistence and robustness against interference, and it uses more bandwidth than the signal transmitted without it. The standard 2.4-GHz band modulation is half-sine filtered offset quadrature phase-shift

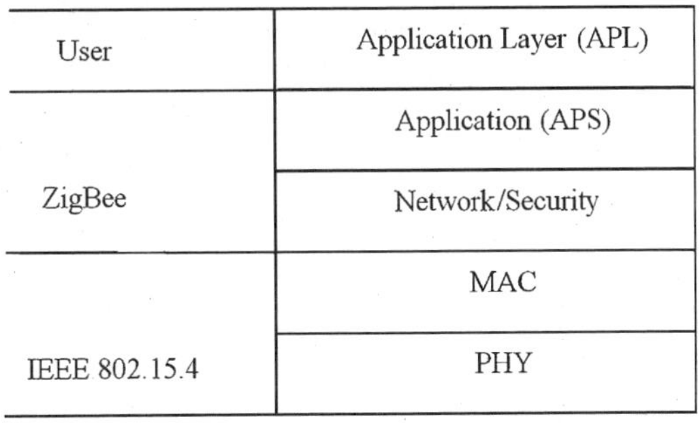

User	Application Layer (APL)
ZigBee	Application (APS)
	Network/Security
	MAC
IEEE 802.15.4	PHY

Fig. 2.2. IEEE 802.15.4 stack

keying (OQPSK) and the 868/915-MHz bands use binary phase-shift keying (BPSK) (Maupin, 2007).

Another technique adopted for coexistence is frequency division multiple access (FDMA), which consists of dividing the 2.4-GHz band into 16 non-overlapping channels with a distance of 5 MHz between them, thus allowing devices operating in adjacent channels to coexist without problems.

In addition to the techniques described previously, Carrier Sense Multiple Access with Collision Avoidance (CSMA/CA) is also necessary in most networks without beacons because several devices may be working in the same channel. This technique basically consists of listening, looking for activity, and, if the channel is busy, waiting a certain amount of time and checking again, and then, if the channel is not busy, using it (ZigBee, 2007; IEEE, 2006).

2.6.2 ZigBee Networks

In addition to the MAC and Physical layers defined by IEEE 802.15.4 mentioned in Section 2.6.1, ZigBee defines two more layers: the Application Support layer (APS) and the Network/Security layer, shown in Fig. 2.2. The user is in charge of the Application layer on top of the Application Support layer.

ZigBee network devices are able to communicate with data rates between 10 and 250 kbps over a 10- to 75-meter range. Depending on the memory requirements and the network size, IEEE short or long addressing can be used (Freescale ZigBee Overview).

Zigbee distinguishes among three types of network devices:

- Full-function device (FFD): FFDs have all the characteristics specified by the IEEE 802.15.4 standard and can work as a network router, end device, or both.
- Zigbee Coordinator (ZC): ZCs store all structure and node information in order to manage the network. For this task to run smoothly, good memory and computing power are fundamental.
- Reduced-function device (RFD): These devices are basically sensors and actuators with limited functionalities like send and/or receive; they know only their function, the location of the ZC, and the nearest router.

ZigBee supports star, mesh, and cluster-tree network topologies. Each has advantages over the others. Star topology networks are the most appropriate for very long battery-life applications and are formed by an FFD working as coordinator and a group of end devices. Mesh topologies provide more than one way through the network in order to increase reliability, and all the available paths are stored in the network routing table. Cluster-tree topologies are a combination of mesh and star topologies as shown in Fig. 2.3; therefore, they have the advantages of both high reliability and long battery life (Freescale ZigBee Overview).

Fig. 2.3. Cluster-tree
topology

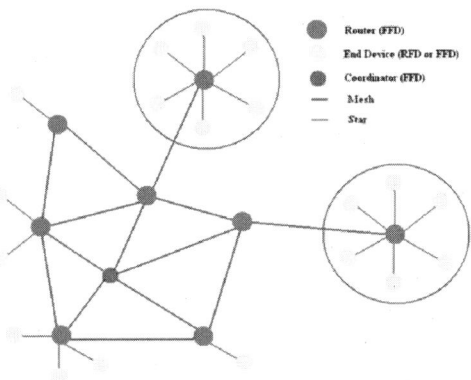

2.6.3 Zigbee Applications

Even though ZigBee is aimed at monitoring and controlling applications, it actually has a much wider range. The applications can be divided into three main groups:

- Nonperiodical applications: These kinds of systems can be developed so that most of their components are disconnected from the network. The devices will connect to the network only when communication is needed, thus consuming very little power. A wide variety of systems fit into this group, such as access control systems, lighting systems, remote control systems, interactive toys, etc.
- Periodical applications: The majority of the systems in this category are measurement systems for applications like patient or fitness monitoring or periodical process control systems. The device must be sleeping and should only wake up at the established time to perform its task; once the task is completed, the device returns to its sleeping mode.
- Periodical low-latency applications: There are also some periodical applications with important requirements like low latency. For these applications, beacon packets and a capability of ZigBee called Guaranteed Time Slots (GTS) need to be used in order to guarantee the time and duration of the communication without time delays (Craig).

2.6.4 ZigBee Promoters and Participants

ZigBee is a profitless alliance of more than 100 companies among semiconductor and electronic device manufacturers, with the goal of promoting this low-cost wireless technology. The ZigBee alliance has three classes of membership: promoters, participants, and adopters.

The promoters consist of semiconductor, software, and system providers and are the leaders of the alliance, representing a cross section of the wireless industry. The promoters of this alliance are BM Group, Ember, Freescale Semiconductor, Honeywell, Huawei, Mitsubishi Electric, Motorola, Philips (Lighting), Samsung, Schneider Electric, Siemens, ST Microelectronics, and Texas Instruments.

Participants play a less important role than promoters in ZigBee. Participants can attend the alliance meetings and have access to all preliminary specifications. Current participants, in alphabetical order, include Ad-Sol Nissin Corp., Airbee, Akita Electronics Systems, Alektrona, AMI Semiconductor, ArchRock, Arcom Solutions, Assa Abloy, Atalum, Atmel, Avocent, Betronic, Bubec, Cambridge Consultants, Certicom, Cirronet, Control 4, Crane Controls Group, Crossbow, Daintree Networks, Danfoss, Develco, Dust Networks, Eaton, eaZix, Eka Systems, Eldatl Embex, Epson, ETRI, Exegin, France Telecom, Fraunhofer, Frontline, Gigatek, Golden Power, Grundfos, Helicomm, Hitachi, Holley, IBBT, Innovative Wireless Technologies, Inovonics, Insta, Invensys, Integrations Associates, Itron, Jenninc, Johnson Controls, KDDI, KETI, Korwin, Legrand, LG, Marlin Controls, MaxStream, MeshNetics, Micrel, Microchip, Mikrokrets AS, Millennial Net, Mindteck, Mono Products, muRata, Nanotron, National ICT, National Instruments, NEC, Nice, Niko, NTS, OKI, Omron, One RF Technology, Orange Logic, OTSL, Radio Pulse, Renesas, RF Technologies, Rincon Research Corporation, Samsung, San Juan Software, SD System, Shinko, Silicon Laboratories, Software Technologies Group, Telecom Lab, Telegesis, Tendril, Trane TR Tech, TSC Systems, TTA, UbiquitousSystem Lab, UBIWave, Urmet Domus, Vantage Controls, Viconics, Vitelec, Winegard, Xanadu Wireless, Yamatake, Yaskawa, Yokogawa, and ZMD.

2.6.5 ZigBee System-on-Chip (SoC)

In most cases, solutions are formed by a transceiver plus a low-power microcontroller, enough to satisfy the wireless sensor's designer's needs. On other applications, a high scale of integration is required. For this purpose, some manufacturers have developed a System-on-Chip (SoC), integrating a microcontroller and a transceiver in the same package. With these integrated circuits (ICs), only a few external passive components are required to build a fully compliant ZigBee device in minimum space. Two examples are Chipcon and Freescale, whose main features are summarized as follows:

- The MC1320X family from Freescale integrates a 2.4-GHz transceiver and a powerful HCS08 processor with several memory combinations in a 5 × 5-mm QFN package (Freescale Home Page; Freescale MC1320x).
- Chipcon also offers a wide variety of RF products and ZigBee solutions such as transceivers, transceivers and microcontrollers with USB, and others

(Chipcon). Interesting products from this manufacturer include the CC2430/CC2431 Systems-on-Chip, which feature an enhanced 8051 microcontroller and an IEEE802.15.4-compliant transceiver (TI, 2006).

2.6.6 Radio-Frequency Integrated Circuit Manufacturers

This section features a compilation of synthesized information from the main integrated circuit manufacturers and from some ZigBee alliance partners that developed embedded systems to evaluate the capacities of the transceivers, microcontrollers, or ZigBee software stacks instead of manufacturing a transceiver. The manufacturers that develop their own transceivers may use evaluation kits to test the effectiveness of their future radio-frequency ICs. The main features of the devices currently considered to be the most innovative and useful are shown below.

2.6.6.1 ZMD

ZMD has a great amount of experience manufacturing application-specific integrated circuits (ZMD Home Page) and focuses on applications for the electronics and automotive industries, medical technologies, and infrared interfaces. ZMD offers the ZM44102 transceiver for ZigBee and more recently developed the ZM44101 System-on-Chip. The advantages of each integrated circuit are described below.

The ZMD44101 is a fully integrated System-on-Chip CMOS transceiver that can operate in the 868.3-MHz European and in the 902-MHz to 928-MHz American ISM (industrial, scientific, and medical) bands. This transceiver has been optimized for low power and to handle data rates up to 40 kbps. In order to ensure reliable data transfers in hostile RF environments, it incorporates DSSS technology. Due to its high scale of integration, the number of external components is minimal (ZMD, 2004).

The ZMD44102 is a robust, fully integrated RF transceiver operating in the license-free European band of 868 to 870 MHz and in the American band of 902 to 928 MHz. ZMD44102 has been optimized for long range and for data rates of up to 40 kbps. DSSS technology is included, enabling ZMD44102 to operate in hostile RF environments (ZMD, 2006).

2.6.6.2 Chipcon

Chipcon, now Texas Instruments, offers a wide range of RF-IC products for short-range, low-power, low-cost, and high-integration wireless applications in the license-free sub-1-GHz and 2.4-GHz bands (Chipcon). The Chipcon ZigBee products are designed for different applications and requirements, such as applications requiring pure transceivers (CC2420), standalone systems (CC2430), or sensor networking (CC2431) (TI, 2006).

The CC2420 is a low-cost, low-power transceiver designed to operate in the 2.4-GHz unlicensed ISM band. It supports AES encryption and authentication, together with cyclic redundancy checksum (CRC) among other features. In addition, we may find the CC2430, a System-on-Chip (SoC) solution for building ZigBee network standard nodes. It has an integrated 8051 microcontroller with a different size of flash memory depending on the version combined with the CC2420 transceiver. Chipcon also has the CC2431 SoC device with similar features to the CC2430, but engine location support is added in order to estimate the node's position in the network based on the signal strength (TI, 2006).

2.6.6.3 Atmel

Like many other silicon manufacturers, Atmel offers solutions for a wide area of wireless applications, including mobile phones, Bluetooth, or WiMAX (Atmel). The AVR-Z link platform is a ZigBee-certified and 802.15.4-compliant solution that allows wireless applications to be easily and quickly developed using free software and development kits. The use of AVR devices as the core of the systems allows the selection of a wide range of microcontrollers that support the porting of the ZigBee stack seamlessly (Atmel AVR). In addition, the AT86RF230 is one of the few transceivers developed by Atmel that is fully compliant with 802.15.4 and ZigBee applications working in the 2.4-GHz band. The AT86RF230 does not need external components to run apart from the antenna, crystal, and decoupling capacitors integrated on the chip, which leads to significant board space saving (Atmel, 2007).

2.6.6.4 Freescale

Freescale is the semiconductors division of Motorola (Freescale Home Page) that has inherited its wide experience in cable and radio communication chips. Freescale focuses on low-power wireless communications and protocols such as ZigBee in the semiconductors market. With this purpose in mind, Freescale has developed several transceiver products, like SoC devices and the MC13xxx family of transceivers (Freescale MC1320x).

The MC13203 is a 2.4-GHz band transceiver designed for wireless sensing and control applications. Like most 802.15.4-compatible devices working in this frequency band, these transceivers support data rates of 250 kbps with O-QPSK modulation and DSSS coding. Its internal architecture combines the reception and transmission circuitry necessary to reduce the number of discrete components needed for normal operation of the transceiver. Like other transceivers, the MC13203 uses the four-wire SPI interface to provide digital data connections with a microcontroller. The user must provide the low-level programming and interfacing to use the 802.15.4 MAC and ZigBee stack supplied by Freescale. This software tool offers compatibility with the MC13xxx family of transceivers and several other microcontrollers from Freescale (Freescale MC13203).

2.6.6.5 Microchip

Microchip is a leader in the microcontroller market because of its PIC devices (Microchip Home Page). At the start of 2007, Microchip released the MRF24J40 (Microchip MRF24J40), its first low-power radio transceiver oriented toward 802.15.4 applications. Along with the release of the transceiver, Microchip announced the MiWi protocol, an 802.15.4-based protocol that is compatible with 802.15.4 devices. The main feature of MiWi is that fewer resources are needed while maintaining compatibility with other ZigBee devices. In addition, it is free, meaning that no royalties or certifications must be paid. This protocol has been designed for low-resource applications in which the number of nodes (1024), coordinators (8), chips per coordinator (127), or hops (4) is critical. One disadvantage is that the stack is designed to be used with Microchip PIC microcontrollers and with the MRF24J40 transceiver (Flowers and Yang, 2007).

Both the ZigBee and the MiWi stacks are fully compatible with the PICDEM Z demonstration board. This board allows easily developing and integrating of ZigBee applications using the Microchip MPLAB IDE as an environment and the MPLAB ICD 2 as an in-circuit programming and debugging tool. The kit includes two PICDEM Z boards with a PIC18LF4620 microcontroller, some push buttons, temperature measurement, I/O interfacing, and a daughter board with the radio chip. The initial versions of this kit prior to the release of the MRF24J40 had a third-party transceiver mounted on the RF daughter board. Finally, the kit includes the ZigBee stack supporting various functionalities (Microchip PICDEM Z).

2.6.6.6 Renesas

Renesas is an emerging corporation that sells silicon devices (Renesas Home Page). Among many integrated circuits, Renesas offers several solutions for the market of ZigBee applications. One of them is the ZigBee Development Kit, which has the necessary items to start building low-power wireless applications immediately. This kit is comprised of the hardware platform, software stack, and development environment. The hardware platform is based on the M16C Renesas microcontroller together with several peripherals like a character LCD, analog, I/O, and radio. The radio is currently only available at 2.4 GHz, but future products will support 900-MHz frequency bands. The software that runs on the M16C microcontroller is a limited version of the ZigBee and 802.15.4 MAC layers. Motes come preprogrammed with an additional demo application that acts as a packet sniffer. This software is useful in order to track packets and data transactions between other nodes when their firmware is being debugged. The development environment is completed with a serial interface that is used to program the platform and to debug user applications through a demo version of the ZigBee Demo Kit ZDK for M16C software (Renesas ZDK).

2.6.6.7 Silicon Laboratories

The IC-specialty firm Silicon Labs offers several solutions for the ZigBee and 802.15.4 protocols (SiLabs Home Page). The microcontrollers of the 8051 family from Silicon Labs are optimal for low-power wireless applications like ZigBee. Silicon Labs has developed microcontroller kits that are available in order to test the platforms and implement future applications. As an example, the ZigBee-2.4-DK is a development kit that allows the user to create ZigBee applications from scratch. It has six target boards and the necessary accessories and software to make it work; each board consists of a Chipcon CC2420 transceiver connected to a C8051F121 microcontroller including USB and JTAG interfaces (SiLabs ZDK; SiLabs AN222). The software tool also consists of an integrated development environment based on the Keil C compiler and software stack, including the ZigBee and 802.15.4 MAC layers. It also includes an application programming interface (API) for personal computers, which has the necessary network primitives to build applications that can manage a ZigBee-based network (SiLabs AN241; SiLabs AN242).

References

Atmel Corporation (2007) ZigBee IEEE 802.15.4 radio transceiver AT86RF230. http://www.atmel.com/dyn/resources/prod_documents/doc5131.pdf.
Atmel Corporation. http://www.atmel.com.
Atmel Corporation. AVR 8-bit RISC. http://www.atmel.com/products/AVR/z-link/Default.asp.
Bluetooth Special Interest Group. http://www.bluetooth.org.
Texas Instruments Inc. http://www.chipcon.com.
Coronis Systems. http://www.coronis.com.
Craig WC. Zigbee: Wireless control that simply works. White paper accessible through http://www.zigbee.org/.
Flowers D, Yang Y (2007) MiWi Wireless Networking Protocol Stack. Microchip Technology Inc. application note.
Freescale Semiconductor, Inc. (2007) Freescale Semiconductor. http://www.freescale.com.
Freescale Semiconductor, Inc. (2007) MC13203: 2.4 GHz RF transceiver for ZigBee applications. http://www.freescale.com/webapp/sps/site/prod_summary.jsp?code=MC13203&nodeId?01J4Fs25658166.
Freescale Semiconductor, Inc. (2005) MC1320x 2.4 GHz RF transceiver. http://www.freescale.com/files/wireless_comm/doc/fact_sheet/MC1320X24GZFS.pdf.
Freescale Semiconductor, Inc. (2007) ZigBee Overview Course. http://www.freescale.com/webapp/sps/site/training_information.jsp?code=WBT_28081&fsrch?1.

IEEE. (2006) IEEE Standard 802.15.4. Telecommunications and information exchange between systems—Local and metropolitan area networks—Specific requirements Part 15.4: Wireless Medium Access Control (MAC) and Physical Layer (PHY) specifications for low-rate wireless personal area networks (WPANs).

Intel Corporation. Ultra-Wideband (UWB) technology. http://www.intel.com/technology/comms/uwb.

Maupin M (2007) An overview of Freescale's ZigBee and IEEE 802.15.4 platforms. http://www.freescale.com/files/ftf_2007/doc/presentations/Americas/Enabling/AE328_FTF2007.pdf?fsrch=1.

Microchip Technology Inc. (2007) Microchip Technology Home page. http://www.microchip.com.

Microchip Technology Inc. (2006) MRF24J40 Data Sheet. http://ww1.microchip.com/downloads/en/DeviceDoc/39776a.pdf.

Microchip Technology Inc. (2007) PICDEM Z Demonstration Kit. http://www.microchip.com/stellent/idcplg?IdcService=SS_GET_PAGE&nodeId=1406&dDocName?en021925.

Nokia Corporation. (2007) Ultra-low power radio technology for small devices. http://www.wibree.com/technology/Wibree_2Pager.pdf.

Renesas Technology Corp. Home page. http://www.renesas.com.

Renesas Technology America, Inc. (2005) ZigBee Demonstration Kit (ZDK) RZB-CC16C-ZDK. http://eu.renesas.com/media/products/connectivity/zigbee/zdk/ZigBeeQuickstart.pdf.

Silicon Laboratories (2006) AN222: 2.4 GHz 802.15.4/ZigBee Development Board Hardware User's Guide. http://www.silabs.com/public/documents/tpub_doc/anote/Microcontrollers/en/an222.pdf.

Silicon Laboratories (2005) AN241: 2.4 GHz ZigBee Network Application Interface Programmer's Guide. http://www.silabs.com/public/documents/tpub_doc/anote/Microcontrollers/en/AN241.pdf.

Silicon Laboratories (2006) AN242: 2.4 GHz ZigBee Network API Programming Example Guide. http://www.silabs.com/public/documents/tpub_doc/anote/Microcontrollers/en/AN242.pdf.

Silicon Laboratories Home page http://www.siliconlaboratories.com.

Silicon Laboratories (2005) 2.4 GHz 802.15.4 Development Kit User's Guide. http://www.silabs.com/public/documents/tpub_doc/evbdsheet/Microcontrollers/en/802.15.4-2.4-DK.pdf.

Texas Instruments Inc. (2006) Low-Power RF Selection Guide. http://focus.ti.com/lit/ml/slab052/slab052.pdf.

Nokia Corporation (2007) Wibree Home page. http://www.wibree.com.

Wi-Fi Alliance Home page. http://www.wi-fi.org.

ZigBee Alliance (2007) ZigBee and wireless radio frequency coexistence. ZigBee white paper.

ZigBee Alliance Home page. http://www.zigbee.org/en/index.asp.

ZMD AG (2004) ZMD44101 single-chip 868 MHz to 928 MHz RF transceiver. http://213.174.55.51/zmd.biz/pdf/ZMD44101_FS.pdf.

ZMD AG (2006) ZMD44102 robust communications RF transceiver 868 MHz to 928 MHz. http://213.174.55.51/zmd.biz/pdf/ZMD44102_Feature_Sheet.PDF.

ZMD AG Home page. http://www.zmd.de.

Chapter 3
Hardware Platforms for WSNs

Abstract Many different manufacturers have complete wireless sensor nodes and solutions available. This chapter describes the current main WSN platforms, ranging from systems with radio-frequency bands of 300 MHz to those platforms using a 2.4-GHz radio-frequency band. Much of the information is presented in tabular form to ease comparison between platforms and to have the data in a uniform format. This information has been compiled from a variety of sources, including product brochures and manufacturers' Web pages. To obtain more detailed information on any product, we highly recommend consulting the references at the end of the chapter or directly contacting the manufacturer.

3.1 AVIDdirector

The American firm AVIDwireless (AVIDwireless Home Page) uses Java in its machine-to-machine (M2M) devices. These devices can be connected to any type of machine, sensor, or device and can transmit information. They are capable of communicating with a great quantity of equipment using wireless networks such as mobile networks, satellites, Wi-Fi, Bluetooth, or ZigBee.

The AVIDdirector device is capable of monitoring the I/O (Input/Output) and sending the information through the different communication channels. These products are based on an "Imsys CJIP Java 160 MIPS" processor that provides a Web interface to manage the options available. The system shows an open architecture, so the final user can build his or her own hardware and use different software platforms. AVIDdirector main characteristics as shown in table 3.1.

3.2 WMSNP

Convergix is a Belgian company specialized in wireless sensor networks and miniature electronic systems (Convergix Home Page). The main WSN product is WMSN.P–2812/D2, which uses a 2.4-GHz ISM band. It is a low-cost, battery-powered module that includes a vibration sensor permitting a wireless sensor network of hundreds of nodes to be built. Table 3.2 summarizes main features of this convergix device.

A.-B. García-Hernando et al., *Problem Solving for Wireless Sensor Networks*,
DOI: 10.1007/978-1-84800-203-6_3, © Springer-Verlag London Limited 2008

Table 3.1 AVIDdirector Features Summary (Compiled from AVIDwireless [AVIDwireless Home page])

Processing Unit		Interface	
Microcontroller	Imsys CJIP Java	Digital	Six GPIO, four high-voltage and current I/O, internal expansion connector
RAM	8 or 16 MB	Analog	Not specified
ROM	4 to 16 MB	Serial	JTAG, Bus I2C, two high-speed serial ports, RS232, TTL serial port
Speed	160 MIPS	Sensor	Temperature
OS	Imsys Java	Infrastructure (gateway)	GSM, GPRS, CDMA, 1xRTT, iDen, Mobitex, DataTac, ReFLEX, Aries, 802.11a/b/g, Bluetooth, ZigBee; RFID and Sirit HF readers (optional)
Radio Frequency		**Miscellaneous**	
Transceiver	Not specified	Power source	12 V DC @ 150 mA
Band	Not specified	Power idle	Not specified
Bit rate	Not specified	Power Rx/Tx	Not specified
Modulation	Not specified	Cost	From $495–1145 depending on the configuration
Range	Not specified	Gateways/mesh	None
Protocol	Not specified	License	Proprietary
Security	Not specified	Extras	Runs Java programs, M2MIO^{TM}, mixed signal array processor

Table 3.2 WMSNP Features Summary (Compiled from Convergix [Convergix Home Page])

Processing Unit		Interface	
Microcontroller	Microchip PIC182220	Digital	None
RAM	512 bytes	Analog	None
ROM	2 kB	Serial	RS232, RS485/422
Speed	8 MHz	Sensor	Acceleration
OS	None	Infrastructure (gateway)	Host computer
Radio Frequency		**Miscellaneous**	
Transceiver	Not specified	Power source	9 V DC (battery)
Band	2.45-GHz ISM band	Power idle	Not specified
Bit rate	9600 bps	Power Rx/Tx	Not specified
Modulation	Not specified	Cost	Not specified
Range	Indoors 50 m, outdoors 2 km	Gateways/mesh	None
Protocol	Not specified	License	Proprietary
Security	Not specified	Extras	256 bytes of EEPROM, ZigBee-compatible

The module is customizable to offer different solutions in order to provide more storage capabilities, different bandwidths, higher transmission rates, and compatibility with other network protocols such as ZigBee.

3.3 SmartMesh-XR

Dust Networks offers wireless sensor networks and low-consumption motes for the 2.4-GHz and 900-MHz bands (Dust Home Page). The company has also developed SmartMesh-XR, a wireless network system that is mote-based using the Time Synchronized Mesh Protocol (TSMP) to provide good network reliability and low power consumption.

SmartMesh-XR has features like redundancy to solve broken links and scalability, enabling up to 250 devices per gateway to be added. Using synchronized network protocols to limit node activity yields low power consumption and reduces the chances of network traffic overload. Network configuration is made easy thanks to the autoconfiguration capability of the nodes, and a Java application is used to manage the network. Acquired data and alarms are transmitted through network motes until they reach the gateway, which manages quality of service and interfaces a personal computer using an XML-based protocol.

This system can be used in typical WSN applications such as weather stations, energy measurement, lightning control, machine state control, process control, equipment control, perimeter monitoring, and intruder detection, among others. Dust Networks' shells offer two families of products depending on the operating frequency band, 2.4 GHz or 900 MHz. The facilities offered by both families are similar, so only the 900-MHz band family will be discussed. In table 3.3 the main characteristics of this family are shown.

Table 3.3 SmartMesh-XR Feature Summary (Compiled from Dust Networks, Inc. [Dust Networks Home Page])

Processing Unit		Interface	
Microcontroller	TSMP engine	Digital	None
RAM	Not specified	Analog	None
ROM	Not specified	Serial	UART 9600 bps
Speed	Not specified	Sensor	None
OS	Not specified	Infrastructure (gateway)	Ethernet, host computer
Radio Frequency		Miscellaneous	
Transceiver	Not specified	Power source	2.7–3.3 V
Band	902–928 MHz	Power idle	8 uA
Bit rate	76.8 kbps	Power Rx/Tx	14 mA/28 mA
Modulation	Binary FSK	Cost	Not specified
Range	200 m outdoors, 80 m indoors	Gateways/mesh	PM1230 gateway
Protocol	TSMP	License	Proprietary
Security	Not specified	Extras	Route-enabled motes

The M1030 mote is a low-power OEM module that allows an easy and reliable way to build a sensor network. It is optimized for the Dust SmartMesh Technology platform and includes a system of wireless sensor with antennas, serial interface, four analog inputs, and four digital input/outputs.

The Manager PM1230 is a network node that provides quality of service and control functions in a mote network. This node also optimizes network topology, manages routes and mote packets, and exports acquired data in XML format through the Ethernet port. An application interface for programming the SmartMesh-XR system from the application is also provided.

3.4 JN5121

Jennic aims to benefit from the advantages that wireless networks give in order to develop new applications for controlling domestic and industrial equipment (Jennic Home Page). There are multiple application fields where these technologies can be implemented such as commercial building for energy saving, industrial process control, and monitoring for high reliability and low maintenance cost, domestic automation, security, and telemetry, among others.

Jennic's JN5121 modules are low-cost, and its low-power microcontroller is IEEE 802.15.4-compliant. Its core supports a RICS 32-bit processor, a 2.4-GHz IEEE 802.15.4-compatible transceiver, 64 KB of ROM, and 96 KB of RAM. The high-scale integration of this device helps to reduce the total cost of the system. For this purpose, ROM provides point-to-point and network protocols and RAM supports routing and function control with no external memory required. Table 3.4 shows a features summary of JN5121 device.

Table 3.4 JN5121 Features Summary (Compiled from Jennic Ltd. [Jennic Home Page])

Processing Unit		Interface	
Microcontroller	32-bit RISC optimized for low power	Digital	21 GPIO
RAM	96 kB RAM	Analog	4 × 12-bit ADC, 2 × 11-bit DAC
ROM	64 kB ROM	Serial	2 × UART, SPI (5 selects), 2-wire
Speed	16 MHz	Sensor	None
OS	Not specified	Infrastructure (gateway)	None
Radio Frequency		Miscellaneous	
Transceiver	Not specified	Power source	2.7–3.6 V
Band	2.4 GHz	Power idle	5 uA
Bit rate	250 kbps	Power Rx/Tx	50 mA/45 mA
Modulation	O-QPSK	Cost	Not specified
Range	> 400 m	Gateways/mesh	None
Protocol	IEEE 802.15.4	License	Proprietary
Security	Security processor (128-bit AES)	Extras	None

The JN5121 platform uses a hardware MAC accelerator and hardware AES encryption accelerator to reduce power consumption and processing. Other energy-saving mechanisms are provided, such as low working time.

The output power is variable in five 6-dB steps, up to 30 dB, with a maximum output power of +2.5 dBm. This is controlled as a result of the RSSI interface with ±1-dB accuracy, 2-dB steps, and ±3-dB linearity. The output power is also not affected by the data rate, which remains constant at 250 kbps as specified by the IEEE 802.15.4 standard.

As this mote is based on the IEEE 802.15.4 standard, the MAC code is not available. Nevertheless, IEEE 802.15.4-compliant devices can handle many more than 100 nodes on a single Previous Access Network Identifier (PANID). The workable maximum number depends on the traffic required over the network.

3.5 MeshScape

Millennial Net offers a wide family of products and services aimed to develop, support, and employ wireless sensor applications, like MeshScape 4.0, a wireless sensor networking system that delivers high scalability, reliability, responsiveness, and power efficiency (Millennial Home Page).

MeshScape 4.0 is available for 916-MHz and 2.4-GHz frequency bands. Main features as shown in Table 3.5. This system is made of hardware control

Table 3.5 MeshScape Features Summary (Compiled from Millennial Net [Millennial Home Page])

Processing Unit		Interface	
Microcontroller	Not specified	Digital	4 × IO (mesh node, end node)
RAM	Not specified	Analog	8 × ADC (mesh node, end node)
ROM	Not specified	Serial	1 serial (mesh node, end node)
Speed	Not specified	Sensor	None
OS	Not specified	Infrastructure (gateway)	RS232 and RS485 (only gateway)
Radio Frequency		Miscellaneous	
Transceiver	Not specified	Power source	CR2032 (end node)
Band	916 MHz, 2.4 GHz	Power idle	Not specified
Bit rate	57 kbps	Power Rx/Tx	Not specified
Modulation	Not specified	Cost	Not specified
Range	20 m (end node), 30 m (mesh node)	Gateways/mesh	Mesh node, MeshGate
Protocol	MeshScape	License	Patented
Security	Not specified	Extras	None

Fig. 3.1. MeshGate module.
(From Millennial Net
[Millennial Home Page].)

network modules like the MeshGate, expansion devices like the meshnodes, and sensor devices like the end nodes, under software-controlled supervision.

As shown in Fig. 3.1, MeshGate is the gateway interface of the MeshScape network. This module configures network parameters, enables networks features, and motorizes the network status.

Figure 3.2 shows a mesh node, which extends network coverage, routes packets in order to avoid obstacles, and provides backup routes in the event

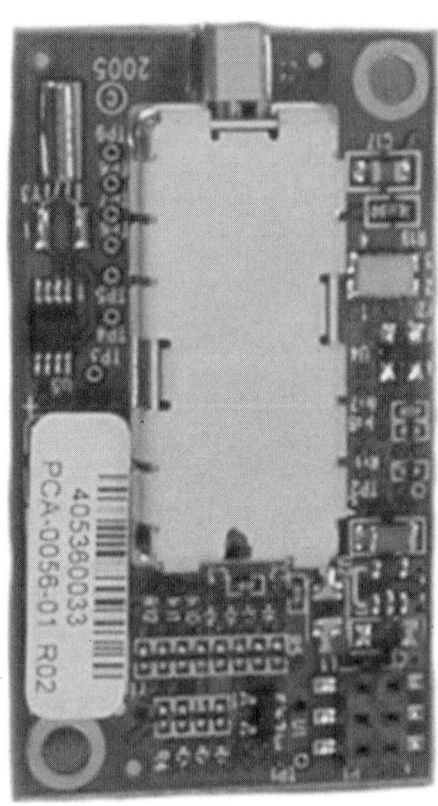

Fig. 3.2. Mesh node. (From
Millennial Net [Millennial
Home Page].)

Fig. 3.3. End node. (From Millennial Net [Millennial Home Page].)

of network congestion or device failure. The mesh node can also be interfaced with sensors and actuators in different network configurations.

End nodes (Fig. 3.3) create an interface between sensors and actuators for acquiring data. These devices work in both a star-network configuration and a star-network hybrid.

3.6 SensiNet

Sensicast offers SensiNet as a complete solution for the 900-MHz and 2.4-GHz wireless sensors (Sensicast Home Page). Specific applications include temperature monitoring, security, industrial monitoring, and building automation. SensiNet relies on two product families for each working band: the H900 group operating at the 900-MHz frequency band and the A2400 group for the 2.4-GHz frequency band.

The H900 network operates in the 900–928-MHz frequency range and uses a slow frequency-hopping spread spectrum with amplified ratios of 12 dBm to provide secure connectivity in point-to-point environments. This system provides a range of 304 meters outside and a range of 91 meters inside. The A2400 network uses IEEE 802.15.4 radios, with the frequency-hopping spread-spectrum ratios amplified to approximately 15 dBm. In this case, the range reduces to 212 meters outside and 70 meters inside.

The SensiNet wireless network is made of smart nodes, mesh nodes, gateways, and bridge nodes. Smart nodes acquire information for the mesh nodes, which then route the information until it reaches the bridge node or gateway. That gateway connects the Sensicast network with other wired networks such as the Internet.

Smart nodes are battery-supplied end devices that communicate with mesh nodes, gateways, or bridge nodes; they sleep most of the time. Sensicast offers a wide variety of sensors, including temperature and humidity sensors.

The SensiNet EMS Smart Sensors are real-time temperature and humidity monitoring devices that are components of the SensiNet product family.

The SensiNet RTD sensor family shown in Table 3.6, is made for outdoor use and is able to stand up to extreme weather conditions. Like the EMS family, it measures humidity and temperature in real time, but it does not have any external connections.

The analog nodes form the SensiNet ALOG family, shown in Table 3.7, which supports voltage and current sensing up to 10 V and 20 mA. With two measurement channels, a 12-bit resolution, and a precision of 0.1%, these modules are suitable for interfacing most analog sensors.

The SensiNet product family also supports security devices, such as close-contact nodes, vibration, and pyroelectric devices. For this purpose, the STAR100 nodes are managed to sense discrete signals like on–off switches,

Table 3.6 SensiNet RTD Models (Compiled from Sensicast Systems [Sensicast Systems Home Page])

	RTD100/110 (H900)	RTD102/4,112/4 (H900)	RTD202/4,212/4 (A2400)
Radio frequency	900 MHz	900 MHz	2.4 GHz
Battery	2/3 lithium	.6 V AA type (lithium thionyl chloride)	.6 V AA type (lithium thionyl chloride)
Autonomy	3 years @ 25 °C with a 1-min transmission interval	RTD102, 3 years @ 25 °C with a 2-min transmission interval RTD104, 1.5 years @ 25 °C with a 2-min transmission interval	RTD102, 3 years @ 25 °C with a 2-min transmission interval RTD104, 1.5 years @ 25 °C with a 2-min transmission interval
Temperature measure range	RTD100 from −100 °C to 200 °C RTD110 from −20 °C to 540 °C	RTD102 from −100 °C to 200 °C RTD112 from −20 °C to 540 °C	RTD102 from −100 °C to 200 °C RTD112 from −20 °C to 540 °C
Processor	Not specified	Not specified	Not specified

Table 3.7 SensiNet's ALOG Models (Compiled from Sensicast Systems [Sensicast Systems Home Page])

	ALOG100	ALOG200	ALOG110	ALOG210
Radio frequency	900 MHz	2.4 GHz	900 MHz	2.4 GHz
Output power	12 dBm	15 dBm	12 dBm	15 dBm
Power	10–30 V DC with battery backup	2/3 AA lithium-ion batteries	2/3 AA lithium-ion batteries	2/3 AA lithium-ion batteries
Input	4–20 mA	4–20 mA	± −10 V	± 0–10 V

and the OAS100 nodes are designed to interface analog sensors such as pressure sensors, vibration sensors, PIR, and magnetic sensors.

Mesh nodes are line-powered nodes with backup batteries that self-organize the network and automatically build routes with other nodes. These mesh nodes also support inputs and outputs for additional sensors. Mesh nodes are wireless intelligent routing modules that receive incoming packets and then route them until they reach the gateway or another mesh node, using several paths to ensure redundancy. Mesh nodes are available for the 900-MHz and 2.4-GHz frequency bands.

Sensicast bridge nodes interface sensor networks with Ethernet-based and serial-based wired networks. Specific software like Sensicast's SensiMesh gateway software is required in order to provide low-level connections and configure sensor networks. On the other hand, the gateways include the SensiMesh software facilities, so no additional software is required.

Sensicast gateways bridge SensiNet networks and cable-based networks such as RS232 and USB. These devices also operate in the 900-MHz and 2.4-GHz frequency bands.

The following table summarizes the properties of the Sensicast product family.

Table 3.8 SensiNet Features Summary (Compiled from Sensicast Systems [Sensicast Systems Home Page])

Processing Unit		Interface	
Microcontroller	Not specified	Digital	None
RAM	Not specified	Analog	Four inputs for voltage and current sensors
ROM	Not specified	Serial	None
Speed	Not specified	Sensor	Temperature, humidity, thermocouple, RTD, contact, voltage, current, vibration
OS	Not specified	Infrastructure (gateway)	Local computer, Ethernet, USB
Radio Frequency		Miscellaneous	
Transceiver	Not specified	Power source	Four AAA-type alkaline batteries for sensors. 4.5–14 V DC and backup batteries for mesh nodes and gateways
Band	900 MHz or 2.4 GHz	Power idle	Not specified
Bit rate	Not specified	Power Rx/Tx	Not specified
Modulation	Not specified	Cost	Not specified
Range	304 m outside, 91 m inside	Gateways/ mesh	Router, SensiNet services gateway
Protocol	SensiNet	License	Proprietary
Security	Not specified	Extras	Compatible with automation software such as Wonderware, LabView, RS View, OSI PI, ICONICS, Citect

3.7 EnRoute

Sensoria's wireless network is based on the EnRoute family (EnRoute500). (On February 1, 2007, Sensoria's EnRoute500 wireless mesh router product family became part of Tranzeo Wireless Technologies USA Inc.) The EnRoute network was originally designed for military purposes and can be used whenever

Table 3.9 EnRoute400 Features Summary (Compiled from Tranzeo Wireless Technologies Inc. [Tranzeo Wireless Technologies Home Page])

Processing Unit		Interface	
Microcontroller	Not specified	Digital	None
RAM	Not specified	Analog	None
ROM	Not specified	Serial	RS232
Speed	Not specified	Sensor	None
OS	Not specified	Infrastructure (gateway)	Ethernet RJ45 connector
Radio Frequency		**Miscellaneous**	
Transceiver	Not specified	Power source	6–16.8 V DC input
Band	2.4–2.483 GHz	Power idle	Not specified
Bit rate	Not specified	Power Rx/Tx	100 mW–2 W
Modulation	Not specified	Cost	Not specified
Range	Not specified	Gateways/mesh	None
Protocol	Not specified	License	Proprietary
Security	AES 128 encryption	Extras	Optional GPS

Table 3.10 EnRoute500 Features Summary (Compiled from Tranzeo Wireless Technologies Inc. [Tranzeo Wireless Technologies Home Page])

Processing Unit		Interface	
Microcontroller	Not specified	Digital	None
RAM	Not specified	Analog	None
ROM	Not specified	Serial	None
Speed	Not specified	Sensor	None
OS	Not specified	Infrastructure (gateway)	10/100 base-T Ethernet, WiFi
Radio Frequency		**Miscellaneous**	
Transceiver	Backhaul connection	Power source	60–280 V AC input
Band	5.15–5.35 GHz, 5.47–5.85 GHz	Power idle	Not specified
Bit rate	6 Mbps, 11 Mbps, 24 Mbps	Power Rx/Tx	12 W
Modulation	BPSK, QPSK, 16-QAM or 64-QAM	Cost	Not specified
Range	Not specified	Gateways/mesh	None
Protocol	802.11a-compatible	License	Proprietary
Security	QoS 802.11e WMM, WPA Ipsec VPN	Extras	Acts as an 802.11b/g access point

reliability and security are compromised. EnRoute devices can be used for perimeter protection, bomb detection, mobile radios, emergency services platforms, CCTV, and VoIP, among others. The EnRoute family of products includes the EnRoute400 and EnRoute500 devices.

The EnRoute400 enables wireless voice, data, and video communication. This device is designed to bridge 802.11b wireless networks with Ethernet and RS232 interfaces with no additional components. It also provides automatic routing connection, intelligent repetition and routing, path redundancy, and VPN for point-to-point communications.

EnRoute500 allows the 802.11a protocol to enable point-to-point communications and the 802.11b/g protocol to act as an access point for WiFi communications. With the same characteristics of EnRoute400, this model integrates a 10/100 Ethernet port and multiple ESSIDs for multiple users. Main features of EnRoute400 and EnRoute500 are shown in Tables 3.9 and 3.10 respectively.

3.8 Tmote Sky

Moteiv offers Tmote Sky as the next generation of low-power motes, with high transmission speed and high fault tolerance for easy sensor network development (Moteiv, 2006).

Tmote Sky was designed at the University of California–Berkeley (USA), by the same developers as for TinyOS. Tmote Sky can be seen in Fig. 3.4 and has been conceived for easy development, programming, and debugging as a result of the USB interface. The integrated antenna allows a range of 125 meters, and the program installed in the 1 MB of flash memory prevents system errors by loading a mirror program on the malfunctioning system.

According to the manufacturer's information, the mote's output power is controllable by the transceiver's microcontroller and power amplifier. The controllable values (32 pre-settable values) allow a range of 5 meters up to 30 or 50 meters indoors, although this can vary with the antenna. In addition, the transceiver's RSSI interface is able to give stable measurements of the output

Fig. 3.4. Tmote Sky. (From [Moteiv, 2006]. Courtesy of Sentilla Corporation.)

Table 3.11 Tmote Sky Features Summary (Compiled from Moteiv [Moteiv Home Page])

Processing Unit		Interface	
Microcontroller	MSP430F1611	Digital	6 GPIO
RAM	10 kB RAM	Analog	6 ADC
ROM	48 kB flash	Serial	I²C, SPI, USB virtual serial com port
Speed	8 MHz	Sensor	Humidity, light, and temperature
OS	TinyOS	Infrastructure (gateway)	Ethernet
Radio Frequency		**Miscellaneous**	
Transceiver	Chipcon CC2420	Power source	Two AA-type batteries (2.1–3.6 V)
Band	2400–2483.5 MHz	Power idle	5.1 uA
Bit rate	250 kbps	Power Rx/Tx	19.5 mA/21.8 mA
Modulation	O-QPSK	Cost	$130/mote ($780/10 motes)
Range	125 m outdoors, 50 m indoors	Gateways/mesh	Tmote Connect
Protocol	802.15.4	License	BSD (operating system)
Security	Those from 802.15.4 protocol	Extras	None

power, which is not affected by the data rate. Table 3.11 shows the main features of Tmote Sky.

Finally, the manufacturer remarks that several different MAC layers are available with the Boomerang and TinyOS 2.x operating systems, both with available source code. A special feature of the CSMA-based MAC layers is that there is no limit on the number of nodes in a single cell.

3.9 MICAx

Crossbow Technology is an important wireless sensor networks manufacturer whose first objective is to use such networks in widely extended commercial applications (Crossbow Home Page). For this purpose, Crossbow provides monitoring systems and automatic acquiring systems that can operate in the 916-MHz frequency band and in the 2.4-GHz band with the Zigbee standard. Crossbow provides a wide variety of hardware and software platforms that give more application development possibilities.

An ATmega 128 processor and an MPR radio module make up the hardware platform devices, also known as motes. The motes are battery-powered devices that form the Crossbow Xmesh network. Each mote ports the Xmesh network software stack and the TinyOS operating system, which enables low-level tasks and events management.

Table 3.12 MICA2 Features Summary (Compiled from Crossbow Technology Inc. [Crossbow Technology Home Page])

Processing Unit		Interface	
Microcontroller	Atmega 128 L	Digital	DIO
RAM	4 kB	Analog	8 ADC
ROM	128 kB flash	Serial	2 UART, I2C, SPI
Speed	8 MHz	Sensor	Humidity, light, and temperature
OS	TinyOS	Infrastructure (gateway)	USB/serial for local computer, or Ethernet at base station
Radio Frequency		**Miscellaneous**	
Transceiver	CC1000	Power source	2 AA-type batteries
Band	868/916 MHz (MPR400CB), 433 MHz (MPR410CB), 315 MHz (MPR420CB)	Power idle	16 uA
Bit rate	38.4 kbaud	Power Rx/Tx	18 mA/35 mA
Modulation	FSK	Cost	$169/unit
Range	Outdoors: 152.40 m (MPR400), 304.8 m (MPR410 and MPR420)	Gateways/mesh	None
Protocol	Not specified	License	BSD (operating system)
Security	Not specified	Extras	512 kB measurement flash

Table 3.13 MICAz Features Summary (Compiled from Crossbow Technology Inc. [Crossbow Technology Home Page])

Processing Unit		Interface	
Microcontroller	Atmega 128 L	Digital	DIO
RAM	4 kB	Analog	8 ADC
ROM	128 kB flash	Serial	2 UART, I2C, SPI
Speed	7.37 MHz	Sensor	Connector for several sensors
OS	TinyOS	Infrastructure (gateway)	USB/serial for local computer, or Ethernet at base station
Radio Frequency		**Miscellaneous**	
Transceiver	CC2420	Power source	2 AA-type batteries
Band	2.4 GHz	Power idle	16 uA
Bit rate	250 kbaud	Power Rx/Tx	27.7 mA/25.4 mA
Modulation	O-QPSK	Cost	$169/unit
Range	75–100 m outdoors, 20–30 m indoors	Gateways/mesh	None
Protocol	802.15.4-compatible	License	BSD (operating system)
Security	AES128	Extras	512 kB measurement flash

Table 3.14 MICA2DOT Features Summary (Compiles from Crossbow Technology Inc. [Crossbow Technology Home Page])

Processing Unit		Interface	
Microcontroller	Atmega 128 L	Digital	DIO
RAM	4 kB	Analog	8 ADC
ROM	128 kB flash	Serial	UART
Speed	4 MHz	Sensor	Connector for several sensors
OS	TinyOS	Infrastructure (gateway)	Serial for local computer, or Ethernet at base station

Radio Frequency		Miscellaneous	
Transceiver	CC1000	Power source	CR2354
Band	868/916 MHz, 433 MHz, 315 MHz	Power idle	16 uA
Bit rate	38.4 kbaud	Power Rx/Tx	18/35 mA
Modulation	FSK	Cost	$142/unit
Range	152.40 m indoors, 304.8 m outdoors	Gateways/ mesh	None
Protocol	Not specified	License	BSD (operating system)
Security	Not specified	Extras	512 kB measurement flash

Crossbow has developed three mote families: MICAz (MPR2400), MICA2 (MPR400), and MICA2DOT (MPR500), which main characteristics are shown in Tables 3.12, 3.13 and 3.14 respectively. MICAz's radio module operates in the ISM 2.4-GHz band and supports the ZigBee standard, Fig 3.6. On the other hand, the MICA2 family, as seen in Fig. 3.5, and the MICA2DOT, Fig 3.7, family operate in the 315-, 433-, and 868-/900-MHz unlicensed frequency bands. These modules are designed for the end user or the OEM product designer.

With respect to development platforms, the MICA platform is a versatile product that has inspired other research projects, like the TelosB mote, which is shown in Fig. 3.8. This project is an open-source platform that joins together

Fig. 3.5. MICA2. (From Crossbow Technology Inc. [Crossbow Technology Home Page]; http://www.xbow.com/Products/productdetails.aspx?sid = 174.)

Fig. 3.6. MICAz. (From Crossbow Technology Inc. [Crossbow Technology Home Page]; http://www.xbow.com/Products/productdetails.aspx?sid = 164.)

Fig. 3.7. MICA2DOT. (From Crossbow Technology Inc. [Crossbow Technology Home Page]; http://www.xbow.com/Products/Product_pdf_files/Wireless_pdf/MICA2DOT_Datasheet.pdf.)

Fig. 3.8 TELOSB. (From Crossbow Technology Inc. [Crossbow Technology Home Page]; http://www.xbow.com/Products/productdetails.aspx?sid = 252.)

essential characteristics for a laboratory environment, such as a USB programming facility, IEEE 802.15.4 radio, low-power MSP430 microcontroller, and sensors. This mote is mainly used in low-power device investigations and wireless sensor network investigations.

Crossbow also developed the CRICKET MOTE (MCS410CA), an MICA2-based module that includes MICA2 hardware and ultrasound transceptors. With these sensors, CRICKET MOTE (shown in Fig. 3.9) is able to measure distances.

Crossbow offers a wide variety of sensors boards that increase application development possibilities and also provides completely mote-compatible external sensors and acquisition boards. Table 3.15 shows the majority of Crossbow's sensors boards.

To interface real-world devices, Crossbow's Gateways and Mote Interface Boards (MIB) enable communications with other devices like PCs and PDAs and interconnections with other networks like the Internet or WiFi.

The MIB family is composed of the MIB510, MIB520, and MIB600 devices, which carry RS232, USB, and Ethernet connectors, respectively.

Fig. 3.9. CRICKET MOTE. (From Crossbow Technology Inc. [Crossbow Technology Home Page]; http://www.xbow.com/Products/productdetails.aspx?sid = 176.)

Table 3.15 Crossbow's Sensors Boards. (Compiled from Crossbow Technology Inc. [Crossbow Technology Home Page])

Sensor	Description
MDA100	Made of a precision thermistor, light sensor, and prototype board
MTTS300/MTS310	Provides a great variety of sensors for MICA, MICA2, and MICAz
MDA500	An acquisition sensor and data board that provides a flexible user interface for external data connection to the MICA2DOT mote
MTS400/420	Provides data acquisition and environment monitoring for MICA2 and MICAz
MTS510	Made for the MICA2DOT board and carries a light sensor, accelerometer, and microphone

The Stargate product family acts as a bridge between mote networks and the Internet. These devices are Linux-based with a CompactFlash, PCMCIA docks, Ethernet, and USB (host) connectors.

Regarding software platforms, MoteWorks is Crossbow's network creation platform for wireless sensor networks. With all these products, Crossbow provides an easy-to-use, reliable solution for OEM developers. Designers should not worry about designing complex hardware and software radio systems for new functionalities. The MoteWorks platform is optimized for battery-powered sensor networks and provides solutions for wireless networks at all levels, including network management support and interface connection.

3.10 BTnodes

Developed at ETH Zurich by the Computer Engineering and Networks Laboratory (TIK) and the Research Group for Distributed Systems, the BTnode, which is shown in Fig. 3.10, is a versatile autonomous wireless communication and computing platform based on a Bluetooth radio, a second

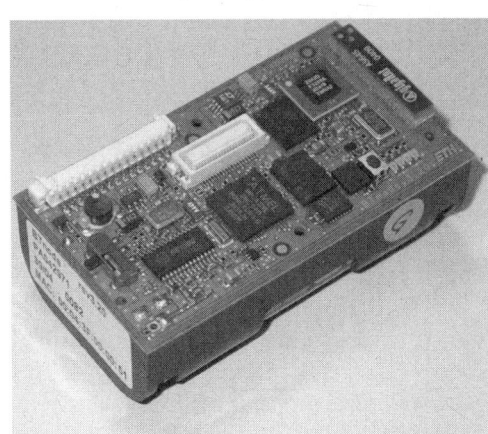

Fig. 3.10. BTnode. (From [BTnodes Home Page].)

Table 3.16 BTnode Features Summary (Compiled from BTnode Home Page])

Processing Unit		Interface	
Microcontroller	Atmega 128 L	Digital	GPIO
RAM	64+180 kB	Analog	ADC
ROM	128 kB flash	Serial	UART,SPI,I2C
Speed	8 MHz	Sensor	Daylight and IR, temperature, microphone, two-axis acceleration sensor
OS	BTnut, TinyOS	Infrastructure (gateway)	None
Radio Frequency		Miscellaneous	
Transceiver	CC1000	Power source	External DC supply 3.8–5 V or 2 AA batteries
Band	868 MHz	Power idle	9.9 mW average
Bit rate	76.8 kbps	Power Rx/Tx	105.6 mW average
Modulation	FSK	Cost	$165/unit
Range	>100 m outdoors with antenna	Gateways/mesh	None
Protocol	Not specified	License	Proprietary
Security	Not specified	Extras	Bluetooth 1.2 Zeevo Radio, BTSense plug-in

low-power radio, and a microcontroller (BTnodes Home Page). It serves as a demonstration and prototyping board for research in mobile and ad-hoc connected networks (MANETs) and distributed sensor networks (WSNs). BTNodes main features are shown in Table 3.16.

The low-power radio is the same as used on the Berkeley MICA2 Motes, making the BTnode a mirror of both the Berkeley MICA2 mote and the old BTnode. Both radios can be operated simultaneously or be independently switched off when not in use, reducing the idle power consumption of the device considerably.

3.11 Embedded Sensor Board

The Embedded Sensor Board (ESB) is a microcontroller-based wireless sensor prototype board equipped with a hybrid transceiver and a set of sensors. This product was developed by a research group (ScatterWeb FuBerlin) at the Freie University of Berlin [FuBerlin Home Page] and is now supported by the spin-off company ScatterWeb.

The core of the ScatterWeb's mote is an MSP430 low-power microcontroller from Texas Instruments, which features 2 kB of RAM memory, 60 kB of flash ROM, and several interface peripherals. The ADC is attached to several sensors that allow measurement of light, passive infrared, vibration, tilt, sound, temperature, etc, Fig 3.11 shows ESB. The integrated radio transceiver allows easy

Fig. 3.11. Embedded Sensor Board. (From Freie Universität Berlin [FuBerlin Home Page].)

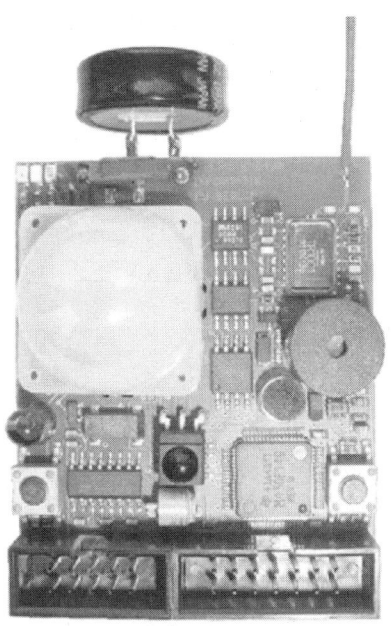

Table 3.17 ESB Features Summary (From Freie Universiät Berlin [FuBerlin Home Page])

Processing Unit		Interface	
Microcontroller	MSP430F149	Digital	8xGPIO
RAM	2 kB	Analog	ADC
ROM	60 kB flash	Serial	JTAG, SPI, I2C
Speed	1 MHz	Sensor	Passive infrared, infrared, temperature, tilt, vibration, sound
OS	Contiki	Infrastructure (gateway)	Serial port for host computer, infrared for PDA or GSM phone; host computer, Ethernet
Radio Frequency		**Miscellaneous**	
Transceiver	TR1001	Power source	Three AAA batteries
Band	868 MHz	Power idle	8 μA
Bit rate	115.2 kbps	Power Rx/Tx	8 mA
Modulation	ASK, OOK	Cost	$149/unit
Range	300 m outdoors, 100 m indoors	Gateways/mesh	eGates
Protocol	Not specified	License	Proprietary
Security	Not specified	Extras	Storage EPROM; Optional Supercap, mains, or solar panel power

communication and an asynchronous serial interface, which, in turn, allows easy integration. Nevertheless, no MAC is integrated in the transceiver, so this task must be done by the microcontroller. On-board devices are powered by three AAA-type batteries, but can also be powered by solar panels, mains, or a Supercap capacitor. Table 3.17 shows the main features of ESB.

3.12 Scattergate and Scatternode

ScatterWeb has proposed a functional system built from nodes and gates (ScatterWeb Home Page). The gate behavior is very versatile, given that they can operate as a gateway or as a mesh node, allowing the network clustering to increase the network size up to 256 devices per gate. The nodes are measure devices compatible with voltage transducers. One advantage of this mote is that it integrates a two-wire serial EEPROM memory for logging purposes. Figure 3.12 shows the ScatterWeb module.

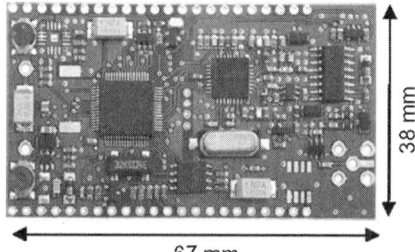

Fig. 3.12. ScatterWeb.
(Courtesy of ScatterWeb
[ScatterWeb Home Page].)

67 mm

Table 3.18 ScatterWeb Features Summary (Compiled from ScatterWeb [ScatterWeb Home Page])

Processing Unit		Interface	
Microcontroller	Not specified	Digital	13 GPIO with interrupt
RAM	Not specified	Analog	A DC
ROM	Not specified	Serial	Not specified
Speed	Not specified	Sensor	Not specified
OS	Not specified	Infrastructure (gateway)	Host computer, Ethernet
Radio Frequency		Miscellaneous	
Transceiver	Chipcon	Power source	1.8–3.3 V (2 AA cells)
Band	868 MHz	Power idle	250 uA
Bit rate	19.2 kbps	Power Rx/Tx	25 mA (Rx)
Modulation	Not specified	Cost	Not specified
Range	1 km outdoors	Gateways/mesh	ScatterGate
Protocol	Not specified	License	Proprietary
Security	Not specified	Extras	RSSI available

3.13 μNodes

WSN platforms developed by Ambient Systems (Ambient Home Page) can be used in a variety of scenarios, such as environmental monitoring, farming, building monitoring, patient monitoring, industrial machines monitoring, logistics, and surveillance. The system is composed of Ambient μNodes, Ambient gateways, and Ambient smart tags. We discuss smart tags in Section 3.14.

System nodes are built from a Texas Instruments MSP430 microcontroller and a 900-MHz-band radio transceiver, both of which ensure low-power operation. The system is able to operate in mesh, star, and hybrid topologies. Ambient uNodes main characteristics are shown in table 3.19.

Table 3.19 μNodes Features Summary (Compiled from Ambient Systems [Ambient Systems Home Page])

Processing Unit		Interface	
Microcontroller	Texas Instruments MSP430	Digital	8 GPIO with interrupt, LCD connector
RAM	10 kB	Analog	8 ADC, 2 DAC
ROM	48 kB	Serial	JTAG, I^{2}C, SPI, RS232
Speed	4.6 MHz	Sensor	Temperature, light, humidity, motion
OS	AmbientRT, TinyOS compatible	Infrastructure (gateway)	LAN, host computer.
Radio Frequency		Miscellaneous	
Transceiver	Not specified	Power source	2 AA batteries, 2.7–3.6 V
Band	868/915 MHz	Power idle	2.5 μA
Bit rate	50 kbps	Power Rx/Tx	12.5 mA/9 mA @ −10 dBm
Modulation	Not specified	Cost	Not specified
Range	50 m indoors, 200 m outdoors	Gateways/mesh	Ambient gateway
Protocol	Not specified	License	Proprietary
Security	Not specified	Extras	4-Mbit onboard storage, 3 onboard LEDs

3.14 Smart Tags

The Ambient smart tag (Ambient Home Page) is a simplified version of the μNodes mote with limited memory, sensing, and actuating capabilities to adjust application requirements, as shown in Smart tags characteristics table; Table 3.20. Smart tags are designed for low-cost and low-power applications and can communicate with the Ambient sensor network. The limitations of the low-power protocol imply that multiple-hop communication cannot be used, meaning that the smart tags are inactive most of time.

Table 3.20 Smart Tags Features Summary (Compiled from Ambient Systems [Ambient Systems Home Page])

Processing Unit		Interface	
Microcontroller	16-bit 8051 microcontroller	Digital	3 GPIO
RAM	Not specified	Analog	None
ROM	Not specified	Serial	I²C, SPI, UART (through previous GPIOs)
Speed	16 MHz	Sensor	Temperature, light, humidity, motion (only one sensor)
OS	None	Infrastructure (gateway)	LAN, host computer
Radio Frequency		**Miscellaneous**	
Transceiver	Not specified	Power source	Coin cell battery
Band	868/915 MHz	Power idle	2.5 μA
Bit rate	50 kbps	Power Rx/Tx	12.5 mA/9 mA @ −10 dBm
Modulation	Not specified	Cost	Not specified
Range	30 m indoors, 100 m outdoors	Gateways/mesh	Ambient gateway
Protocol	Not specified	License	Proprietary
Security	Not specified	Extras	32 kbit onboard EEPROM storage, one onboard LED and switch

3.15 Wavecard, Waveflow, Wavetherm, Wavesense, and Wavefront

Wavenis offers a family of radio-frequency products that cover many wireless applications. This family is based on the Waveflow, Wavesense, Wavetherm, Waveport, OEM Wavecard, and Wavefront products, each intended for industrial or commercial purposes. This platform excludes the HCI and microcontroller stack, although it supports the software stack.

The Wavenis protocol stack offers tree, star, and meshed network topologies as well as an innovative self-configuring and self-routing algorithm dedicated to ULP networks. Wavenis products can also work in point-to-point, broadcast, polling, and repeater modes.

3.15.1 The Wavecard and Waveport Platforms

The Coronis Wavecard, shown in Fig. 3.13, is suitable in applications where fast integration of wireless features is necessary but RF hardware customization is not desired. Fully functional out-of-the-box, Wavecard includes the ASIC Wavenis RF transceiver and protocol stack in a single unit that can be plugged directly into the assembly or motherboard. Table 3.21 shows the main features of Wavecard.

Fig. 3.13. Coronis
Wavecard

Table 3.21 Coronis Wavecard Features Summary

Processing Unit		Interface	
Microcontroller	TI MSP430F149	Digital	4 GPIO
RAM	2 kB RAM	Analog	None
ROM	64 kB flash	Serial	RS232 or I²C interface for application motherboard connection
Speed	4 MHz	Sensor	Internal temperature
OS	None in standard	Infrastructure (gateway)	LAN, host computer
Radio Frequency		**Miscellaneous**	
Transceiver	ASIC Wavenis	Power source	Main power or 1 AA battery, 3.6 V/3200 mA
Band	868 MHz/915 MHz/ 433 MHz	Power idle	3 μA
Bit rate	2.4–100 kbps (typically 20 kbps)	Power Rx/Tx	18 mA/45 mA @ +15 dBm10 μA @ 1-s latency
Modulation	GFSK	Cost	OEM Wavecard: 60 € /unit Waveport plastic box: - RS232: 150 €/unit, - USB: 130 €/unit, - CompactFlash: 300 €/unit
Range	25 mW: 1 km (outdoors)/200 m (indoors) 500 mW: 5 km(outdoors)/ 800 m (indoors)	Gateways/ mesh	Self-organizing, self-healing
Protocol	Wavenis	License	Proprietary
Security	FHSS, data interleaving, forward error correction, GFSK modulation	Extras	128-kbit onboard EEPROM storage

Fig. 3.14. Coronis
Waveports

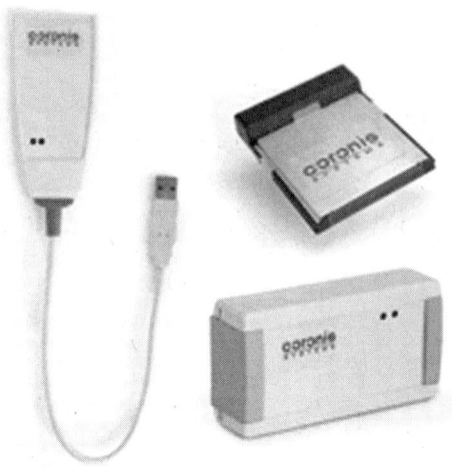

The Waveport, Fig. 3.14, (which has the Wavecard inside) USB, RS232, or CompactFlash products are suitable in applications where using a host computer interface is necessary.

3.15.2 The Wavesense, Wavetherm, and Waveflow Platforms

Coronis' Wavesense, Wavetherm, and Waveflow platforms are suitable in applications where it is necessary to quickly integrate wireless features with specific sensors (such as analog, temperature, or pulse), but where RF hardware customization is not desired, Fig 3.15. Fully functional out-of-the-box, the Wavenis RF transceiver and protocol stack come with specified sensors in a single unit. Table 3.22 shows its main features.

Fig. 3.15. Coronis' Wavesense, Wavetherm, and
Waveflow

Table 3.22 Coronis' Wavesense Features Summary

Processing Unit		Interface	
Microcontroller	TI MSP430F149	Digital	8 GPIO
RAM	2 kB RAM	Analog	ADC 12 bits
ROM	64 kB flash	Serial	None
Speed	4 MHz	Sensor	Temperature, pulse, humidity, analog (0–5 V/ 4–20 mA)
OS	None in standard	Infrastructure (gateway)	LAN, host computer
Radio Frequency		**Miscellaneous**	
Transceiver	ASIC Wavenis	Power source	1 AA battery, 3.6 V/ 3200 mA
Band	868 MHz/915 MHz/ 433 MHz	Power idle	3 µA
Bit rate	2.4–100 kbps (typically 20 kbps)	Power Rx/Tx	18 mA/45 mA @ +15 dBm 10 µA @ 1-s latency
Modulation	GFSK	Cost	Waveflow OEM: 35€ /unit, Waveflow pack: 70€ /unit, Wavetherm OEM: 45€ / unit, Wavetherm pack: 120€ / unit, Wavesense OEM: 45€ /unit, Wavesense pack: 70€ /unit,
Range	25 mW: 1 km (outdoors)/200 m (indoors) 500 mW: 5 km (outdoors)/800 m (indoors)	Gateways/ mesh	Mesh network: self-organizing, self-healing
Protocol	Wavenis	License	Proprietary
Security	FHSS, data interleaving, forward error correction, GFSK modulation	Extras	128-kbit onboard EEPROM storage

3.15.3 The Wavefront Platform

The Coronis Wavefront solution allows the user to build wireless-enabled products with a single microcontroller for the Wavenis protocol stack and its applications, Fig 3.16. The Wavenis PDK gives the user full control over all stack and RF features in order to enhance the customer's own value-added services. The main features of wavefront are summarized in table 3.23.

Fig. 3.16. Coronis
Wavefront

Table 3.23 Coronis' Wavefront Features Summary

Processing Unit		Interface	
Microcontroller	(connect to the Wavefront) TI MSP430F149	Digital	8 GPIO
	Microchip PIC16 and PIC18 NXP and ST (ARM7 TDMI)		
RAM	Depends on the μC	Analog	None
ROM	Depends on the μC	Serial	SPI
Speed	Depends on the μC	Sensor	None
OS	Depends on the μC	Infrastructure (gateway)	OEM card
Radio Frequency		**Miscellaneous**	
Transceiver	ASIC Wavenis	Power source	Not specified
Band	868 MHz/915 MHz/433 MHz	Power idle	2 μA
Bit rate	2.4–100 kbps (typically 10 kbps)	Power Rx/Tx	18 mA/45 mA @ +12 dBm
			10 μA @ 1-s latency
Modulation	GFSK	Cost	Wavefront OEM: 15€/ unit
Range	25 mW: 1 km (outdoors)/200 m (indoors)	Gateways/ mesh	Mesh network: self-organizing, self-healing
	500 mW: 5 km (outdoors)/800 m (indoors)		
Protocol	Wavenis	License	Proprietary
Security	FHSS, data interleaving, forward error correction, GFSK modulation	Extras	1-kbit onboard EEPROM storage

3.16 eyesIFX

The eyesIFX mote is a product developed by Infineon in collaboration with
the Technical University of Berlin and other European partners, under the
project EYES. It is based on one of the most powerful processors in the

Table 3.24 eyesIFX Features Summary

Processing Unit		Interface	
Microcontroller	MSP430F1611	Digital	DIO
RAM	10 kB	Analog	ADC
ROM	48 kB	Serial	SPI
Speed	8 MHz	Sensor	Temperature, light, RSSI
OS	TinyOS	Infrastructure (gateway)	None
Radio Frequency		Miscellaneous	
Transceiver	TDA5250	Power source	Coin cell battery
Band	868 MHz	Power idle	0.2 mA
Bit rate	19.2 kbps	Power Rx/Tx	9 mA/12 mA
Modulation	FSK	Cost	Kit with five motes: ∼300 €
Range	15–30 m indoors	Gateways/mesh	None
Protocol	Not specified	License	Proprietary
Security	Not specified	Extras	Serial data flash 4 Mbit

MSP430 family, combined with a transceiver from Infineon and a USB interface. Some sensors and digital outputs are present on the board. The main advantage of this platform lies in the Tiny OS operating system, which, in conjunction with the nesC and MSP430 libraries, makes eyesIFX a good development platform. Table 3.24 shows eyesIFX main features.

3.17 WSN Platforms' Comparative

Tables 3.25 to 3.30 show the parameters of the platforms described above in a format that facilitates their comparison.

Table 3.25 WSN Platforms' Comparative—Part 1

Parameter\ Platform	AVIDdirector	WMSN.P	SmartMesh M1030	JN5121
Develop, debug, and support	Yes	No	No	Yes
Simulation or profiling tools	Yes	Yes	Yes	Not specified
Programming	JTAG	Not available	Not available	Serial
Programmability through network	Not specified	Not available	Not specified	Not specified
Sensor interface	6 DIO, 4 high-voltage and current I/O, internal expansion connector	Acceleration	LVTTL UART	21 DIO; 4 ADC, 2 DAC; 2 UART, SPI, 2-wire

Table 3.25 (continued)

Parameter\Platform	AVIDdirector	WMSN.P	SmartMesh M1030	JN5121
Programming language	Java	Not available	Not available	C
Operating system	Not specified	None	Not specified	Not specified
Number of nodes	Not specified	Hundreds of sensors	250 per manager	Not specified
Consumption	12 V DC @ 150 mA	Not specified	Power idle 8 µA Power Rx/Tx 14 mA/28 mA	Power idle 5µA Power Rx/Tx 50 mA / 45 mA
Range	Not specified	50 m indoors, 2 km outdoors	200 m outdoors, 80 m indoors	> 400 m
Processor	Imsys CJIP Java	Microchip PIC182220	TSMP engine	32-bit RISC optimized for low power
Bit rate	Not specified	9600 bps	76.8 kbps	250 kbps
Added-value parameters	Data privacy and security; real-time data processing	Up to seven independent networks; true point-to-point; retry and ack; seven hopping channels.	Encryption, key exchange; multiple networks; time synchronized mesh protocol	AES128; security and encryption
Enclosure	Plastic box	Plastic box	None	None
Cost	From $495 to $1145 depending upon configuration	Not specified	Not specified	Not specified

Table 3.26 WSN Platforms' Comparative—Part 2

Parameter\Platform	MeshScape	SensiNet	EnRoute400	EnRoute500
Develop, debug, and support	No	No	No	No
Simulation or profiling tools	Yes	Yes	Yes	Yes
Programming	Not available	Not available	Not available	Not available
Programmability through network	Not available	Not available	Not available	Not available
Sensor Interface	4 DIO (mesh node, end node); 8 ADC (mesh node, end node); 1 serial (mesh node, end node)	Temperature, humidity, thermocouple, RTD, contact, voltage, current, vibration	RS232	None
Programming language	Not available	Not available	Not available	Not available
Operating system	Not specified	Not specified	Not specified	Not specified
Number of nodes	Hundreds of nodes	Not specified	Not specified	Not specified

Table 3.26 (continued)

Parameter\Platform	MeshScape	SensiNet	EnRoute400	EnRoute500
Consumption	Not specified	Not specified	Power Rx/Tx 100 mW–2 W	Power Rx/Tx 12 W
Range	20 m (end node), 30 m (mesh node)	304 m outdoors, 91 m indoors	Not specified	Not specified
Processor	Not specified	Not specified	Not specified	Not specified
Bit rate	57 kbps	Not specified	Not specified	6 Mbps, 11 Mbps, 24 Mbps
Added value parameters	Hop-by-hop packet acknowledgment	Not specified	Security and encryption; VPN; traffic prioritization	QoS 802.11e WMM, WPA Ipsec VPN
Enclosure	None	Plastic box	Plastic box	Plastic box
Cost	Not specified	Not specified	Not specified	Not specified

Table 3.27 WSN Platforms' Comparative—Part 3

Parameter\Platform	Tmote Sky	MICA2	MICAZ	MICA2dot
Develop, debug, and support	Yes	Yes	Yes	Yes
Simulation or profiling tools	Not specified	Yes	Yes	Yes
Programming	USB, JTAG	Serial, JTAG, USB, Ethernet	Serial, JTAG, USB, Ethernet	Serial, JTAG, Ethernet
Programmability through network	Yes	Yes	Yes	Yes
Sensor Interface	6 DIO; 6 ADC; I^2C, SPI, USB virtual serial com port; humidity, light, and temperature	DIO; 8 ADC; 2 UART, I2C, SPI	DIO; 8 ADC; 2UART, I2C, SPI	DIO; 8 ADC; UART
Programming language	nesC	nesC	nesC	nesC
Operating system	TinyOS, Boomerang Contiki-ompatible	TinyOS	TinyOS	TinyOS
Number of nodes	Not specified	Thousands	Thousands	Thousands
Consumption	Power idle 5.1 uA Power Rx/Tx 19.5 mA/ 21.8 mA	Power idle 16 uA Power Rx/Tx 18 mA/ 35 mA	Power idle 16 uA Power Rx/Tx 27.7 mA/ 25.4 mA	Power idle 16 uA Power Rx/Tx 18/ 35 mA
Range	125 m outdoors, 50 m indoors	Outdoors: 152.4 m (MPR400), 304.8 m (MPR410 and MPR420)	75–100 m outdoors, 20–30 mcindoors	152.4 m indoors, 304.8 m outdoors
Processor	MSP430F1611	Atmega 128 L	Atmega 128 L	Atmega 128 L

Table 3.27 (continued)

Parameter\Platform	Tmote Sky	MICA2	MICAZ	MICA2dot
Bit rate	250 kbps	38.4 kbaud	250 kbaud	38.4 kbaud
Added-value parameters	Hardware link-layer encryption and authentication	Not specified	AES128 encryption	Not specified
Enclosure	None	None	None	None
Cost	$130/mote ($780/ 10 motes)	$169/unit	$169/unit	$142/unit

Table 3.28 WSN Platforms' Comparative—Part 4

Parameter\ Platform	BTnode	Embedded Sensor Board	Scatternode	µNodes
Develop, debug, and support	Yes	Yes	No	Yes
Simulation or profiling tools	Not specified	Yes	Yes	Yes
Programming	Serial, JTAG	JTAG	Not available	JTAG,
Programmability through network	Yes, firmware update	Not specified	Not available	Yes, applications, tasks, and business
Sensor Interface	Interface DIO; ADC; UART, SPI, I2C; daylight and IR; temperature, microphone, 2-axis acceleration sensor	8 DIO; ADC; SPI, I2C; passive infrared, infrared, temperature, tilt, vibration, sound	13 DIO with interrupt; 5 ADC, 1 DAC	8 DIO with interrupt, LCD connector; 8 ADC, 2 DAC; I²C, SPI, RS232; temperature, light, humidity, motion
Programming language	C	C	Not available	C
Operating system	BTnut, TinyOS compatible	Contiki	Not specified	AmbientRT, TinyOS-compatible
Number of nodes	Not specified	Not specified	256 nodes per Scatter Gate	Not specified
Consumption	Power idle 9.9 mW average Power Rx/Tx 105.6 mW average	Power idle 8 µA Power Rx/Tx 8 mA	Power idle 250 uA Power Rx/Tx 25 mA (Rx)	Power idle 2.5 µA Power Rx/Tx 12.5 mA/9 mA
Range	>100 m outdoors with antenna	300 m outdoors, 100 m indoors	1 km outdoors	50 m indoors, 200 m outdoors
Processor	Atmega 128 L	MSP430F149	Not specified	MSP430
Bit rate	76.8 kbps	115.2 kbps	19.2 kbps	50 kbps
Added value parameters	Not specified	Not specified	Not specified	Frequency hopping
Enclosure	None	None	None	None
Cost	165€/unit	149€/unit	Not specified	Not specified

Table 3.29 WSN Platforms' Comparative—Part 5

Parameter\Platform	Smart Tags	Wavecard/Waveport	Wavesense/Wavetherm/ Waveflow
Develop, debug, and support	Yes	Yes	Yes
Simulation or profiling tools	Yes	Not available	Not available
Programming	Serial	JTAG	JTAG
Programmability through network	Yes, applications, tasks, and business	Not available	Not available
Sensor interface	3 DIO; I^{2}C, SPI, UART (through previous GPIOs); temperature, light, humidity, motion (only one sensor)	4 GPIO; I^{2}C, SPI, UART (through previous GPIOs); internal temperature	8 GPIO; temperature, pulse, humidity, analog (0–5 V/4–20 mA)
Programming language	C	Assembler, C	Assembler, C
Operating system	None	None	None
Number of nodes	Not specified	Thousands	Thousands
Consumption	Power idle 2.5 µA Power Rx/Tx 12.5 mA/9 mA	18 mA/45 mA @ 15 dBm 10 µA @ 1-s latency	18 mA/45 mA @ 15 dBm 10 µA @ 1-s latency
Range	30 m indoors, 100 m outdoors	25 mW: 1 km (outdoors), 200 m (indoors) 500 mW: 5 km (outdoors), 800 m (indoors)	25 mW: 1 km (outdoors), 200 m (indoors) 500 mW: 5 km (outdoors), 800 m (indoors)
Processor	16-bit 8051	TI MSP430F149	TI MSP430F149
Bit rate	50 kbps	2.4 kbps to 100 kbps (10 kbps typ)	2.4–100 kbps (typically 10 kbps)
Added-value parameters	Frequency hopping	FHSS, data interleaving, forward error correction, GFSK modulation, RSSI	FHSS, data interleaving, forward error correction, GFSK modulation, RSSI
Enclosure	None	Plastic box or none	Plastic box or none
Cost	Not specified	From 60 to 300€/unit (depending upon configuration)	From 35 to 120€/unit (depending upon configuration)

Table 3.30 WSN Platforms' Comparative—Part 6

Parameter\Platform	Wavefront	EyesIFX
Develop, debug, and support	Yes	Yes
Simulation or profiling tools	Not available	Not specified
Programming	Depends on µC	USB, JTAG
Programmability through network	Not available	Not specified
Sensor interface	8 GPIO; SPI (through previous GPIOs)	SPI, DIO, ADC; temperature, light, RSSI
Programming language	Assembler, C (depend on µC)	nesC
Operating system	Nucleus, CMX or None	TinyOS

Table 3.30 (continued)

Parameter\Platform	Wavefront	EyesIFX
Number of nodes	Thousands	Not specified
Consumption	18 mA/45 mA @ 12 dBm	Power idle: 0.2 mA, power
	10 µA @ 1-s latency	RX/Tx: 9 mA, 12 mA
Range	25 mW: 1 km (outdoors), 200 m (indoors)	15–30 m indoors
	500 mW: 5 km (outdoors), 800 m (indoors)	
Processor	TI MSP430F149 Microchip PIC16, PIC18	MSP430F1611
	NXP and ST (ARM7)	
Bit rate	2.4–100 kbps (typically 10 kbps)	19.2 kbps
Added-value parameters	FHSS, data interleaving, forward error correction, GFSK modulation, RSSI	None
Enclosure	None	None
Cost	From 5 to 15€/unit (depending upon quantities)	300€ /kit with five motes

3.18 Open Issues in Hardware Platforms for WSNs

Several issues need to be improved in terms of hardware platforms for WSNs. Perhaps the most important issues in order to increase popular use and production are size and cost. In order to disseminate this kind of technology, cost is an important factor, given that one WSN can use hundreds of sensors, making every little increase in price have an exponential effect on the total price.

The boom in the use of wireless sensor networks has caused different chip manufacturers to develop new circuits adequately integrated for these types of applications. This fact is quite relevant and will make it possible in the near future for the development of wireless applications to increase considerably. The future development of this technology will integrate more functions and reduce the size. In this regard, the contribution from silicon manufacturers will be very important for reducing price and size.

New encapsulation formats and monochip solutions can help to reduce size with better production quality. Custom solutions or FPGA solutions can also help pave the way. With these two factors, more general uses will be available to meet customer needs.

Another important point is that the hardware should have the capacity to work with very small voltages and very small power consumption, in order to reduce the size of the batteries and to prolong their life. However, power consumption cannot be reduced only by using lower voltage; a better design of the MAC components, hardware, and software must also be implemented. Other concerns regarding the MAC components include reducing consumption, maintaining latency, delivering data, and meeting real-time constraints.

Reducing consumption consists of two issues: extending the WSN applications while reducing the price of the final solution. More battery changes are needed if consumption is high, which will have a direct effect on the final price. Reducing consumption by half means reducing the number of battery changes by half, which implies savings in both maintenance and batteries.

In addition, in order to further increase capacity and mitigate the impairment by fading, delay-spread, and co-channel interference, multiple-antenna systems have been used for wireless communications. Such systems have been proposed for point-to-multipoint one-hop cellular networks.

Antenna diversity is based on the fact that signals received from uncorrelated antennas have independent fading. It is thus highly probable that the receiver can get at least one good signal (Rappaport, 2002).

Antenna correlation is usually achieved through space, polarization, or pattern diversity. The processing technologies for diversity include switch diversity, equal gain, and maximum ratio combining. When strong interference is present, diversity processing alone is insufficient to receive high-quality signals. To solve this issue, adaptive antenna array processing is used to shape the antenna beamform so as to enhance the desired signals while nullifying the interfering signals.

The technique for adaptive antenna processing is called optimum combining. It assumes that part of the desired signal information can be acquired through a training sequence. Antenna diversity and smart antenna techniques are also applicable to WSN and other ad hoc networks (Ye et al., 2002).

Due to complexity and cost, a fully adaptive smart antenna system is only used in the base stations of cellular networks. Ongoing research and development efforts are still needed to implement a fully adaptive smart antenna system in a mobile terminal. For WSNs, low cost is a challenging issue. Consequently, directional antennas have been actively researched in the area of ad hoc networks.

A mechanically or electronically steerable or switched directional antenna system can be tuned to a certain direction. By using directional transmission, interference between network nodes can be reduced, thus improving network capacity. A directional antenna can also improve energy efficiency, but it leads to other challenges to the MAC protocol design.

For directional and smart antennas, many MAC protocols have been proposed for ad hoc networks. However, for multiple-antenna systems, an efficient MAC protocol to attain significant throughput improvement is still needed. Communication protocols for cognitive radios remain an open issue, and significant research efforts are needed to make cognitive radio-based WSNs practical.

Another open issue in order to improve reliability is embedding some aspects such as security or QoS in the hardware. This could be a concern when dealing with frequency hopping or encryption methods that avoid eavesdropping or improve QoS depending on the final requirements (real time and latency). Perhaps future developments in the area of wireless sensor networks will necessitate this kind of integration of chip solutions.

Due to the ever-increasing access to technology, security and QoS have clearly been proven to be of the utmost importance. Although previously less of a concern, they are obviously essential in today's world of wireless technology.

References

Ambient Systems Home page. http://www.ambient-systems.net.

AVIDwireless Home page. http://www.avidwireless.com.

BTnode Project. BTnodes—A distributed environment for prototyping ad hoc networks. http://www.btnode.ethz.ch.

Convergix Home page. http://www.convergent-electronics.com.

Crossbow Technology Inc. Home page. http://www.xbow.com.

Dust Networks, Inc. Home page. http://www.dustnetworks.com.

Tranzeo Wireless Technologies Inc. EnRoute500 Series. http://www.tranzeo.com/products/radios/EnRoute500-Series.

Freie Universität Berlin Home page. http://www.fu-berlin.de/en.

Jennic Ltd. Home page. http://www.jennic.com.

Millennial Net Home page. http://www.millennialnet.com.

Moteiv Home page. http://www.moteiv.com. (Note: The new page is Sentilla Corporation. http://www.sentilla.com.)

Rappaport TS (2002) *Wireless Communications: Principles and Practice*. Prentice Hall PTR, Upper Saddle River, NJ.

Freie Universität Berlin. ScatterWeb. http://cst.mi.fu-berlin.de/projects/ScatterWeb.

ScatterWeb Home page. http://www.scatterweb.com.

Sensicast Systems Home page. http://www.sensicast.com.

Ye W, Heidemann J, Estrin D (2002) An energy-efficient MAC protocol for wireless sensor networks. *In Proceedings of the 21st Annual Joint Conference of the IEEE Computer and Communications Societies (INFOCOM 2002)*, Vol. 3, pp. 1567–1576.

Chapter 4
Software Technologies in WSNs

Abstract WSNs pose many challenges to the software applications that run on them. It is not only the scarcity of nodes' resources, but also the particularities of the WSN applications that make software design and development in WSN an open research issue. This chapter reviews some of the most significant software-related aspects of WSNs that are currently the object of intensive research, namely middleware for WSNs, the applicability of agent technologies to WSNs, and design strategies for and the operation of WSN software. We also review WSN simulation platforms. Although they are not software to be run on top of a WSN, these platforms are a key element when devising new protocols or solutions for this technology, as they eliminate the need to deploy a real WSN network from the beginning.

4.1 Middleware Architectures for WSNs

Middleware is considered a necessary layer among the hardware, operating systems, batteries, and application. The aim of the middleware layer is to provide

- Appropriate interfaces to diverse applications
- A run-time environment that supports and coordinates multiple applications
- Mechanisms to achieve adaptive and efficient use of system resources

Middleware has often been used by traditional systems as a bridge between the operating systems and the applications. It makes the development of distributed applications possible. Traditional distributed middleware (DCOM, CORBA) is not adequate for the WSN requirements of memory, energy, and computation. The maintenance of traditional middleware architectures is also not easy due to WSN constraints. For these kinds of networks, a middleware that is simple, light, and easy to implement is needed.

Recent investigations on middleware architectures aim to facilitate high-level WSN programming and optimize resources in order to increase the growth of sensor networks. Some investigators support middleware as way of linking the application with the set of network protocols. This middleware layer must provide an abstraction of portable and standardized systems, support the coordination of competing applications, and facilitate the development of WSN applications.

4.1.1 Characteristics of WSN Middleware

WSN middleware should support the implementation and basic operation of a sensor network while taking into consideration some of the unique characteristics of WSNs:

- Sensor nodes are small-scale devices (with volumes approaching a cubic millimeter in the near future).
- Sensors are limited in the amount of energy stored and/or harvested from the environment.
- Sensors are likely to fail, due to depleted batteries or to environmental influences.
- Sensors have restricted resources (CPU performance, memory, wireless communication bandwidth and range).

Node mobility, node failures, and environmental obstructions cause frequent network topology changes. Communication failures are also a typical problem in wireless sensor networks. Another issue is heterogeneity since the network may consist of a large number of rather different nodes in terms of sensors, computing power, and memory. On the one hand, the large number raises scalability issues; on the other hand, it provides a high level of redundancy. Nodes also have to be able to operate in unattended mode since it is impossible to service a large number of nodes in remote or inaccessible locations.

In order to deal with the characteristics outlined above, WSN middleware must face the following challenges:

- Supporting the development, maintenance, deployment, and execution of sensing-based applications. This includes mechanisms for defining complex, high-level sensing tasks, communicating these tasks to the WSN, coordinating sensor nodes to split and distribute the tasks to each node, gathering data to merge the sensor readings of the individual sensor nodes into a high-level result, and reporting the results back to the task issuer.
- Working in a network with a great number of wirelessly connected nodes (sensors).
- Providing abstraction of the network for heterogeneity among the different components of the WSN.
- Fulfilling the main requirements of WSNs, namely, energy efficiency, reliability, and scalability.
- Allowing event-based or periodic communications. These approaches represent the characteristics of the WSN better than the traditional scheme (based on requests and responses).

- Providing support for automatic configuration and fault management, which are necessary for the unattended way the nodes operate.
- Paying attention to the concepts of time and location. These are the key elements for unifying the information obtained by the different sensors.
- Providing application knowledge in nodes. Middleware for WSNs has to provide mechanisms for injecting application knowledge into the infrastructure and the WSN.
- Providing support for real-time applications as needed.

4.1.2 Various Middleware WSN Approaches

Different middleware approaches were selected and classified taking the programming models used into account.

Programming sensor networks includes two major classes (see Fig. 4.1). The first one is programming support, which manages the providing systems, services, and run-time mechanisms, such as reliable code distribution, safe code execution, and application-specific services. The second one is programming abstraction, which is related to the way a sensor network is viewed and presents concepts and ideas of sensor nodes and sensor data.

4.1.2.1 Programming Support

The programming support class consists of five approaches (see Fig. 4.1): virtual machine–based, modular programming–based, database-based, application-driven, and message-oriented middleware.

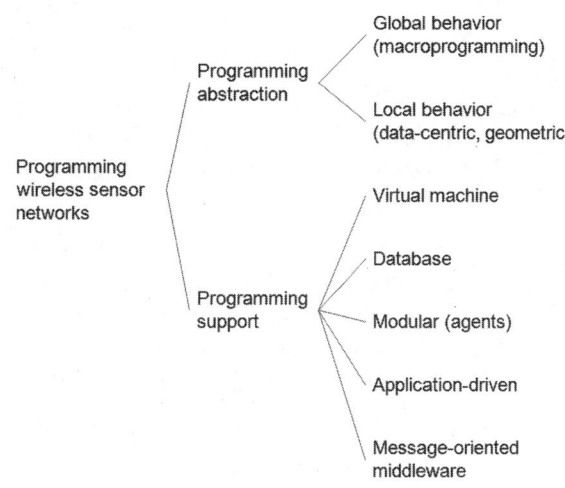

Fig. 4.1 Middleware approaches taking the programming model used into account. (From [Hadim and Mohamed, 2006]. © 2006 IEEE.)

Virtual Machine

This approach consists of virtual machines (VM), interpreters, and mobile agents. Its main characteristic is flexibility, allowing developers to write applications in divided small modules, which are injected and distributed through the network by the system using tailored algorithms and then interpreted by the VM. Those tailored algorithms minimize the overall energy expenditure as well as resource use. However, the technology is complex and the instructions introduce overhead.

- **Maté** (Levis and Culler, 2002) offers improved interaction and adaptation to the variability of sensor networks by supporting openness and scalability. Network protocols and parameters are updated by active messages as a consequence of injecting a new module. As Maté's programs are both short and failure-resistant, the network is dynamic, flexible, and easily reconfigurable. The use of different ad hoc routing protocols and protocol updates deals with mobility. Nevertheless, the instruction interpretation overhead makes Maté inefficient for complex applications, reducing its field of operation to low-duty cycle applications. As of this writing, Maté is simply architecture and byte code, making it difficult to use and requiring further development.
- **Squawk** (Simon et al., 2006) is a virtual machine (VM) written mostly in Java that runs without an operating system on a wireless sensor platform. Squawk provides a wireless API that allows developers to write applications for WSNs. This API is an extension of the generic connection framework (GCF). Squawk enables the authentication of deployed files on the wireless device and the migration of applications between devices. The main problem is that the Squawk VM is mainly applied to the Sun Small Programmable Object Technology (SPOT) wireless device (SunSpot) (a device developed at Sun Microsystems Laboratories to experiment with wireless sensor and actuator applications).

Modular Programming (Mobile Agents)

The use of mobile code facilitates the injection and distribution through the network and leads to application modularity. Less energy is necessary when broadcasting small modules instead of the complete application.

- **Impala** (Liu and Martonosi, 2003) makes use of an appropriate architecture model that provides application adaptation at run time and offers security against inopportune programming errors. Mobility, openness, and scalability are supported by changing between different protocols and modes of operation depending on the applications and network conditions. Impala uses a new autonomic method to select and change to adequate protocols. The maximum modularity leads to higher energy efficiency for sensor node applications, supporting updates of small mobile agents that generate small transmission overhead and energy expenditure. Nevertheless, Impala is predetermined to execute only on Hewlett-Packard/Compaq iPAQ Pocket PC handhelds running Linux and does not support heterogeneity in terms of hardware platforms. Thus, its applications are restricted and it is not directly applicable to WSNs.

- The **Smart Messages Project** (Smart Messages) also makes a proposal consisting of a distributed model called "Cooperative Computing" in which migratory units called smart messages are defined.

Database

This approach observes the entire network as a virtual database system, offering an easy-to-use interface that permits the user to extract data of interest and issue queries about the sensor network. Nevertheless, this approach does not support real-time applications, as it provides only approximate results and the detection of spatial-temporal relationships between events is not possible.

- **Cougar** (COUGAR) establishes an innovative dimension in middleware research by assuming a database approach in which sensor data form a virtual relational database. WSN management operations are executed in the form of queries by means of an SQL-like language. Cougar characterizes a sensor database system that includes a sensor database containing stored data, sensor data, and sensor queries. Stored data are represented as relations consisting of a collection of sensors that contribute to the sensor database and the sensor characteristics. Sensor data are generated by signal processing functions and are then characterized as time series to make sensor queries easier to generate. The system uses abstract data types including virtual relations to model the signal processing functions, which it represents as sequences and supports, making use of incremental results to preserve a constant observation of long running queries. In order to reduce the energy expenditure necessary to collect the information and estimate the result of a request, the power is controlled by the distribution of requests between nodes.
- **TinyDB** (Madden et al., 2003) is a query-processing system for removing information from a network of sensor devices. It uses TinyOS as an operating system, which usually obliges the user to write embedded C code to extract sensor data. However, TinyDB simplifies the extraction of sensor data by providing an SQL-like interface. To specify the type of readings and the subset of nodes of interest, the queries use simple data manipulation. For this reason, TinyDB keeps a virtual database table with columns containing information regarding sensor types, sensor node identifiers, and remaining battery power. In order to launch the queries throughout the network, the system makes use of a controlled-flooding approach. TinyDB keeps a routing tree with a root at the endpoint, which is usually the user's physical position. Consequently, in a decentralized approach, every sensor node includes its own query processor to process and collect sensor data and to keep all routing information. For data collecting, the parent node closer to

the root agrees with the children nodes. The entire process is done again for each period and query.

- **SINA** (Shen et al., 2001) stands for "System Information Networking Architecture" and models the network as substantially distributed objects. The system is based on clusters, and its core is a database for querying and monitoring. The total sensor network is a set of datasheets. Each logical datasheet contains cells that characterize sensor node attributes. Each cell is exclusive and each sensor node keeps the entire datasheet. The database approach facilitates information management by assembling application modifications and requirements. The system supports scalability and energy saving via hierarchical clustering and manages the database by an associative broadcast that leads to an attribute-based naming scheme. The cells start in a node by SQL-like queries from other nodes. The nodes use four different approaches to keep the cells: on-demand content retrieval, content coaching, periodic content update, and triggered content update.
- **DSWare** (Li et al., 2003) is a database approach based on event detection. The system supports flexibility by means of group-based decision making and reliable data-centric storage. DSWare also supports real-time applications and reduces transmission overheads. DSWare provides an SQL-like interface for registering and cancelling events.

Application-Driven

This approach establishes a new, innovative aspect in middleware research by complementing an architecture that accomplishes the network protocol stack, enabling programmers to adjust the network according to the exact application requirements. It provides a QoS advantage since the applications determine the network operations management.

- **Milan** (Murphy and Heinzelman, 2002), which stands for "Middleware Linking Applications and Networks," is used on applications that affect the whole network. The applications' quality needs can be specified and the network characteristics adjusted to extend the application lifetime while still meeting those needs. Milan selects the correct combination of sensors that assures application QoS requirements, making use of specialized graphs that include state-based changes in application needs. Milan can then configure and manage the network with its architecture extending into the network protocols stack and an abstraction layer that lets network-specific plug-ins convert commands to protocol-specific commands. The set of sensor nodes that best assemble the requirements is established by the network plug-ins. The system combines the two restrictions to obtain a general set of possible combinations.

Message-Oriented Middleware (MOM)

This approach is essentially a communication model in a distributed-sensor network. The system facilitates message exchange between nodes and the sink nodes by means of a publish-subscribe mechanism. This model supports asynchronous communication, making movable combinations between the sender and receiver possible.

- **MIRES** (Souto et al., 2004) implements a component-based programming model using active messages to put its publish-subscribe–based communication infrastructure into practice. It includes (1) a core component, which is a publish-subscribe service, that synchronizes the communication between middleware services, (2) a routing component, and (3) a data aggregation service that allows the user to indicate how data will be gathered and to specify the association between sensed data of interest and aggregates. This system supports asynchronous communication by means of its core component, is built on TinyOS, and uses embedded C code. Communication takes place in three phases: First, the network nodes announce their sensed data through the publish service. Next, the system routes the presented messages to the sink, using the multi-hop routing algorithm. Finally, the user application subscribes to sensed data of interest using an appropriate GUI. The system sends only messages referring to subscribed sensed data, thereby decreasing the number of transmissions and the energy expenditure.
- **SensorBus** (Ribeiro et al., 2005) is a message-oriented middleware model for WSNs based on the publish-subscribe paradigm. It allows the free exchange of the communication mechanism among sensor nodes, which, in turn, allows more than one communication mechanism to address the requirements of a larger number of applications.

4.1.2.2 Other Programming Support Approaches

Some other middleware designs also give programming support yet are not so easily classifiable into one of the categories listed above. Several of these efforts are summarized in the following paragraphs.

AutoSec

AutoSec (Han and Venkatasubramanian, 2001) stands for "Automatic Service Composition." This application-driven middleware supports dynamic service brokering, which leads to better efficiency in terms of the use of resources in distributed systems. AutoSec manages the network's resources by providing the QoS necessary for these distributed applications by selecting both information-collection and resource-provisioning strategies from a given set related to user and system needs.

Agilla

(Fok et al., 2005). Based on Maté, Agilla offers improved mechanisms for inserting a mobile code into the sensor network in order to manage user applications. Mobile agents are able to move themselves into desired locations, thus adapting to network changes.

Garnet

Garnet (St Ville and Dickman, 2003) manages data streams as an abstraction in a sensor network. It provides a set of system services such as receivers, filtering and dispatching services, resource manager, and orphanage.

(Yu et al., 2004)

We include a proposal by Yu et al. in which certain design principles that motivate a framework based on clusters are exposed. Their approach provides a virtual machine that separates the application semantics of the infrastructure.

In general, a cluster is a set of spatially adjacent sensor nodes that reside around the target phenomena and are capable of detecting and/or processing the data of interest. Clusters are dynamically formed during the lifetime of the system, triggered by the changing conditions of the environment, data source, and sensor nodes. Each cluster contains some contiguous nodes that cooperate as a functional middleware basic unit. Yu et al. (2004) deal with the topics related to the implementation of middleware for a wireless sensor network with a cluster-based architecture (see Fig. 4.2).

The abstraction created by the middleware is called a virtual machine because it provides semantics that are transparent to the application's physical infrastructure. As Fig. 4.2 shows, the middleware architecture is divided into two layers:

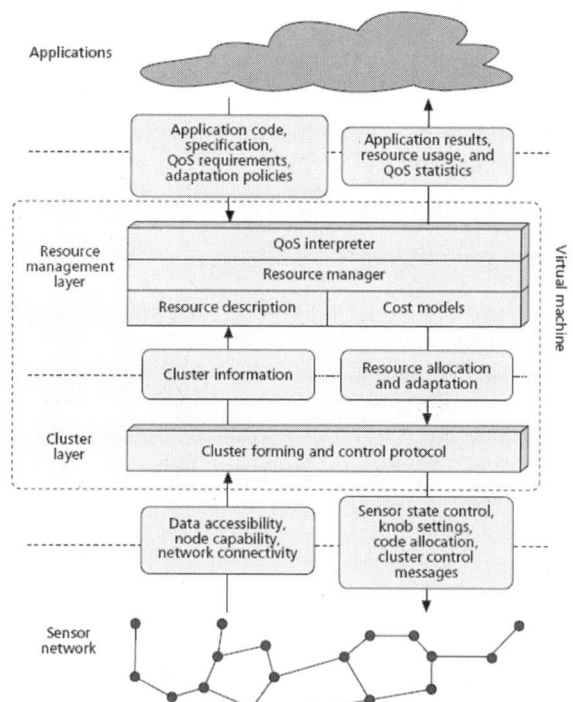

Fig. 4.2 Architecture for a cluster-based middleware. (From [Yu et al., 2004]. ©2004 IEEE.)

- Cluster layer: It forms a cluster with the nodes surrounding the target.
- Resource management layer: The main component of the middleware, this layer arranges the assignment and adjustment of resources in order to satisfy the application's QoS requirements.

Other elements of this middleware include:

- Cluster control: It is necessary to develop self-configuring, on-the-fly distributed mechanisms of clustering.
- Resource management: Gathering and updating the information are key in order to reach an agreement between the system's energy consumption and response time.
- Inter-cluster coordination: It is important to establish mechanisms to detect the existence of a cluster alias, which is a node that belongs to two or more clusters, and to coordinate the elements that form the cluster.

Yu et al. (2004) considered the following characteristics in the design of the middleware:

- The middleware must provide data-centric mechanisms for data processing and must use a cluster-based network model that is robust, simple, and flexible.
- Application knowledge: Knowledge is useful for software design and implementation and should be integrated with the middleware-provided services. A good policy is to integrate the application characteristics inside its code or specification, which the middleware can interpret.
- It is probable that not all of the application's run-time requirements are satisfied, due to limited resources.

(Sharifi et al., 2006)

This work, entitled "A middleware layer mechanism for QoS support in wireless sensor networks," proposes a real-time and fault-tolerant mechanism as a middleware possibility for WSNs that operates according to sensor nodes' requirements such as data rate and energy. A service-based middleware receives the users' QoS requirements about wireless sensor network services and guarantees time-critical responses that are both efficient and scalable in a cluster-based organization.

4.1.2.3 Programming Abstractions

There are two main approaches for programming abstraction classes (see Fig. 4.1): the global behavior (or macro-programming) and the local behavior approaches.

Global Behavior

This first programming abstraction approach, also called macro-programming, revolves around viewing the global behavior of a distributed sensor network as a whole. This method introduces a completely different approach on how to program sensor networks. With this approach, the sensor network is programmed as a whole rather than writing low-level software to drive individual nodes. A global WSN's behavior is programmed at a high-level specification that enables node behaviors to be automatically generated. Application developers do not need to be concerned about dealing with low-level software for each network node. Some examples of this approach follow.

- **Kairos** (Gummadi et al., 2005) provides the necessary notions and concepts in order to design, develop, and implement a macro-programming model on a WSN. Developers using Kairos can express a single centralized program (global behavior) in subprograms that can be executed on local nodes (nodal behavior). This includes compile- and run-time subsystems and leaves the programmer only a small set of programming primitives. The Kairos system makes most of the low-level concerns such as distributed code generation, remote data access and management, and internode program flow coordination clear to the programmer. Kairos provides three main abstractions to a programming language: manipulation of nodes through node abstraction, tracking a current list of node radio neighbors through a list of one-hop neighbors, and reading from variables at named nodes through a data access mechanism. With Kairos it is possible to choose the address node synchronization option: loose synchronization or tight synchronization. The programmer has to decide between efficiency and system overhead.
- Other research projects have adopted the macro-programming models, among them **Regiment** (Newton and Welsh, 2004), a functional system driven by demand that views the sensor network as a whole, **Abstract Task Graph** (Bakshi et al., 2005), a data-driven programming model system that incorporates extensions for distributed sense-and-respond applications, and **Semantic Streams** (Whitehouse et al., 2005), a mark-up and declarative query language system based on using semantic services accessed directly by query specifications.

Local Behavior

This second programming abstraction approach deals with the behavior of the sensor network nodes from a local point of view in a distributed computation. The local behavior approach focuses on the nature of the sensed data and, in particular, on a specific location in a sensor network. An example can be an application that requests a particular location where the temperature is above a certain value rather than asking individual sensors for their readings. The following paragraphs summarize several local behavior proposals.

- **Abstract regions** (Welsh and Mainland, 2004) are a group of general-purpose communication primitives for WSNs that provide addressing, in-network data aggregation, data sharing, and data reduction in the local regions of the network. Local computation is increased and radio communication is reduced, which serves as an efficient method for energy and bandwidth savings. This type of system makes use of a group of nodes that cooperate and communicate locally where the computation and aggregation of data take place. For example, in a military environment, tracking a mobile target involves gathering all sensor readings from nodes near the object to generate accurate information on the target's characteristics. The best approach for an application trying to determine the boundaries of a region of interest in a network is to have nearby nodes cooperate and compute the boundaries from a local point of view. The results are sent to the base station, which concentrates on collecting the final data from each sensor. Each sensor only has to communicate to the base station to send the final collected data, which saves energy as there is less communication overhead.

- **EnviroTrack (data-centric)** (Abdelzaher et al., 2004) makes use of a data-centric programming paradigm called attributed-based naming through context labels. In this paradigm, the routing and addressing are based on the requested data content rather than on the target sensor node's identity. This system is suited for embedded tracking applications, is built on top of TinyOS, and uses compiled nesC programs. Programmers apply the attributed-based naming by associating user-defined entities, or context labels, to actual physical targets The dynamic behavior of tracked targets, such as mobility, is supported in this system. Thus, the system is useful for environmental monitoring and military applications because it detects and reports the presence of any moving target. EnviroTrack is a middleware service among a whole set of other middleware services carried out at the University of Virginia under a major initiative called VigilNet, which is an integrated sensor network system for surveillance missions.

- Other projects that adopt a local nodal behavior approach include **Hood** (Madden et al., 2002), which shows neighborhood abstraction of sensor nodes that can communicate local behaviors, and **generic role assignment** (Römer et al., 2004), which allows programmers to assign individual nodes to user-defined roles.

4.2 Agent Technologies for WSN

4.2.1 Agent Technology and Models

4.2.1.1 General Characteristics of Agent-Based Systems

The term "agent" has a great variety of definitions, including

- "An agent is everything that could be seen perceiving its environment across sensors and acting towards the environment across a few actuators" (Russell and Norvig, 1995).
- "Intelligent agents realize three functions: to perceive environment dynamical conditions, to act concerning the environment conditions and to interpret the perceived information. They resolve problems, show interfaces and determine actions" (Hayes-Roth, 1995).

According to these definitions, it is possible to consider agents as entities inside an environment prepared to feel and act, and therefore qualified to communicate and collaborate with other entities, either humans or other agents. Agents provide a uniform syntax and play a semantically consistent intermediate role.

Taking into account their attributes, in a particular environment we can distinguish different kinds of agents:

- Intelligent agents: entities that emulate mental processes or simulate rational behavior.
- Personal assistants: entities that help users perform a task.
- Mobile agents: entities capable of moving across a network to obtain their goals. They are capable of migrating their code between nodes.
- Information agents: agents that filter and organize coherently dispersed and unrelated data.
- Autonomous agents: agents capable of performing unsupervised actions.

According to the agents' characteristics, we can define a multi-agent system as a distributed system formed by agents capable of cooperating with each other and sharing resources in order to complete a common or individual task. For that reason, they must be capable of interacting in a common environment, while having communication, negotiation, and coordination abilities. The aim of a multi-agent system is to become an autonomous system.

The main characteristics of a traditional program that differ from those of an agent-based program consist of the following pair:

- Autonomy: Agents make their own decisions since they are directed by their own aims.
- Every agent has its own execution thread. Nevertheless, a traditional program has only one flow for the whole system.

Some traditional applications of multi-agent systems are as follows:

- Web finders: Several agents that carry out functions similar to those of an "electronic secretary," capable of actuating in representation of its user and requesting necessary information from other agents.
- Server systems: One or more agents that represent the system in the network. Those agents are capable of talking with other agents, providing them with information about the characteristics of the server and the data contained therein. When a change takes place, only the information relating to the change in the server agent needs to be introduced, and then this agent informs the others of the change.
- Industrial applications such as manufacturing, air-traffic control, and process control.
- Commercial applications such as

    - Information administration: To reduce the amount of data, agents filter the data, search in databases, and show useful information to users.

– Electronic trade: Agents can look in the network for the products that may be of interest to a user.
– Medical applications such as patient monitoring.

4.2.1.2 Agent Platforms

Agents need certain platforms that provide them with a series of services such as identification, mobility, communication, and others for their execution. We briefly describe some proposals of multi-agent system architectures from different workgroups.

KAoS

KAoS stands for "Knowledgeable Agent-oriented System" and proposes an open architecture that describes agent implementations and the communication between them, which is carried out by messages.

OMG

The most important agent model from the OMG (Object Management Group) is the MASIF (Mobile Agents Standard Interface Facility) created for the development of static and mobile agents. The agents are characterized for their capabilities, type of interaction, and mobility.

FIPA

The protocols created by FIPA (IEEE Foundation for Intelligent Physical Agents) are the most used worldwide protocols for agent development, management, naming, and location. FIPA has developed specifications that allow multi-agent system development to be carried out based on "a minimal frame for the agent management on an opened environment" (www.fipa.org) by using

- A reference model that specifies the environment where the agents exist and operate
- An agent platform that provides an infrastructure for the deployment and interaction of the agents

FIPA standards can be classified in three groups:

- Component: specifications that provide the standardization of agent-related technologies
- Informative: presents different applications in a specific context
- Profiles: component specifications to verify that the different implementations agree with the standards

FIPA has developed its own agent platforms corresponding to its specifications. One of them is JADE (Java Agent Development Framework), which was created to simplify the development of distributed agent-based applications. JADE is Java-base and allows interoperability with other agent platforms.

LEAP (Lightweight Extensible Agents Platform) continues with the philosophy of the JADE platform from which it was derived, since it was actually developed as an API executed in JADE. The LEAP project allows agent integration in Java (J2ME) using the FIPA standards. LEAP is developed for limited devices like PDAs or cell phones. This platform provides a modular architecture with one part common to all the devices and another part to be integrated depending on the specific device.

4.2.2 Use of Agent Models in WSNs

4.2.2.1 Similarities Between Mote and Agent Models

An agent can be considered an autonomous program that detects any occurrence or event taking place in its environment and then taking steps or acting to reach its objectives for the designated environment.

An agent receives perceptions or information from the environment from a few sensors. These perceptions will be processed by its software, which will decide which actions must be carried out and order those actions to be performed by the actuators. The agent model presents several analogies with the usual sensor model that appears in the literature (shown in Fig. 4.3). As the figure shows, motes are composed of sensors and a process unit. The sensors obtain information from the environment, while the process unit manages the environmental information obtained from other motes or from an intruder and gives the order to the actuators. The communications module makes it possible to communicate with other motes to obtain information or to perform a specific action.

The agent model seems to have a natural application in the mote field, at least for its components. In fact, Lesser et al. (2003) proposed a sensor network

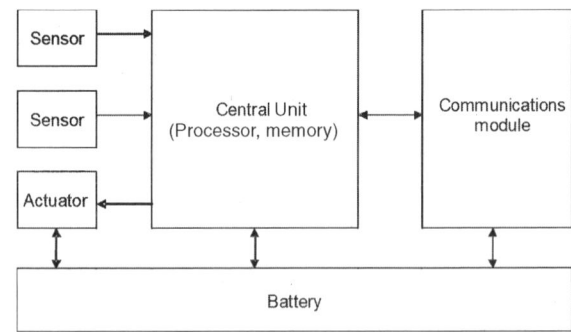

Fig. 4.3 Model of a sensor. (From [Blumenthal et al., 2003]. © 2003 IEEE.)

application consisting of an agent model in which every sensor is controlled by an agent.

The battery is included explicitly as one of the main components in the sensor model. It is the energy source, which is very valuable because in many cases its reload or change is not possible, allowing three types of operations to be carried out by the device: measurement/actuation, processing, and communication. The strong restrictions in energy consumption on which the sensors have to operate may be critical in terms of some of the usual characteristics of the traditional agent environment. For example, the mobility of the agents, meaning the transfer of its code from one device to another, may mean an inadmissible waste of energy according to the energy-saving policies.

4.2.2.2 Application of Agent Models' Characteristics in WSNs

The following agent characteristics can be applied in sensor network environments:

- Reactivity: Agents are capable of detecting and reacting to stimuli. They are capable of processing the data they receive from the sensors and responding to these data.
- Asynchronous work capacity: Agents start to work when one or more sensors detect an event, such as an intruder in the network using a perimeter security.
- Autonomy: Agents have a certain type of control over their actions. They detect a problem, decide what action to take, and solve conflicts. This is important in WSNs because if one element fails, the agent has to be capable of reaching its goal without this element.
- Objective orientation: Agents are capable of handling a task to reach their objective. This is logical due to the fact that they must handle the information provided by the sensors in order to perform their tasks.
- Communication capacity: Agents must be capable of interacting with other agents, sensors, and even humans since the purpose is to provide information or a solution to one or more stimuli.
- Collaboration and cooperation: Agents must be capable of carrying out their tasks by means of cooperation with other agents. In fact, they must provide and obtain information from other elements, thus forming a distributed environment in order to solve certain problems.

In summary, the applicability of agent technology to WSNs comes from the similarity between their objectives. Agents are entities with certain autonomy that respond to events and try to reach concrete goals. This is very similar to what happens in some applications in sensor networks. A clear example is intruder detection: When an intruder is detected, the application adapts to control it, warn the appropriate personnel, and track the intruder's movements by means of cooperation among different nodes of the network.

Mote networks can be organized by means of a mesh or cluster topology. Cluster topologies are more common. In a cluster topology, every sensor is associated and managed by an agent. Some of the advantages of this type of organization include

- Motes are limited in energy, storage capacity, and processing ability. For this reason, only the parts involved in the operation need to be functioning, which implies saving power. Thus, when agents are used, only the corresponding agents in the nodes near the events consume resources. If the agents are mobile as well, they can move to other nodes depending on the dynamics of the objective.
- The interaction between sensors and therefore the processes of the agents are minimized, since one centralizes and reduces the information flow. In every cluster there is an element in charge of information aggregation of the nodes that belong to the cluster.
- It allows the dynamic adjustment of agents to the network: When a change takes place in a particular sector, it does not necessarily have to concern the other sectors. For example, it is possible to modify the infrastructure of the network inside a building which does not affect any other area. This also facilitates the network's scalability.

4.2.2.3 Indications for Using Agent Models in WSNs

Agent technology was not initially intended as an application for wireless sensor networks, since WSNs have strong resource limitations. Agents can be defined as entities in a certain environment where they can feel, act, communicate, and collaborate with other entities. Several authors working on the application of agent technology in mote networks have proposed innovative variations on the classical agent model such as

- The size and the resources consumed by agents in the nodes must be reduced, especially if they are mobile agents that waste energy when communicating with each other.
- The energy used in communications is higher than the energy consumed by data processing inside the nodes. Therefore, one of the agent's main missions should be to complete a preprocessing, aggregation, or filtering of the information before its transmission. The aim is to make the volume of information transmitted as small as possible.
- The classical agent environments (FIPA or OMG) are too heavy for mote platforms. Therefore, it is necessary to consider specializing these devices for specific environments.
- As sensors must perform various tasks, there is a need to use different kinds of agents, some of which can be very sophisticated. In these cases, it is necessary to have a hierarchy among the elements of the network and to know the agents' different tasks. Therefore, agents needing a lot of resources can run inside nodes with high processing power.

4.2.3 Specific Proposals Applicable to WSNs

Certain examples have been selected using the proposed agent application in WSNs to improve the safety of a specific area such as intruder detection or target tracking in a monitoring area. Some of the characteristics described generally in previous paragraphs can be observed in concrete systems.

4.2.3.1 Application with Static Agents

According to a proposal presented by Lesser et al. (2003), the network is organized in clusters with a certain number of sensors by sector. The usual number is eight sensors per sector, although 10 is considered to be a maximum and five sensors a minimum. Each agent must know the size of its sector (e.g., rectangular sectors) since the agents are in charge of determining the behavior of the objects. Agents must know whether or not some object belongs to their sector.

Inside each sector, agents are specialized to perform tracking and network exploration tasks. Every sector has a sector manager that generates and distributes the needed plans to control new targets, assigns roles to different agents, and stores and provides local sensor information as part of the directory service. The manager also acts as a hub, facilitating data interchanges between interested parts as well as changes in the agent population.

The sector manager stores a list with the local agents acting as track managers and stores estimations of the tracked targets as well. The track manager tracks the target. In order to perform this task, it must be capable of selecting the sensors that provide more reliable measurements in each moment. In order to know a target's exact position, information from three or more sensors is needed.

It is possible that the selected sensors do not provide correct measurements or provide information relating to a different target. The sector manager must know if this information is from a new or previously detected target. To do this, every track manager must provide the sector manager with continuous information about tracked targets.

When selecting the tracking agent, the sector manager uses several criteria:

- Work estimations of every agent are needed to avoid overloading any one agent.
- Maintain a minimum of one cluster alias in the channel.
- Geographical location; due to real-time requirements, the nearer, the better.
- Experienced agents that have previously performed this role are preferred.

As soon as a target is assigned to the track manager, the tracker must request directory service information from the sector manager. The manager will then return all the available sensors in order to select the suitable nodes and update its local directory service to avoid future requests, thereby saving energy.

The target can enter and leave different sectors. Two mechanisms support this:

- Migration: If the target and the needed sensors are too far, the track manager informs the sector manager, which then contacts a second sector manager or selects a more suitable tracking sensor.
- AOI (area of influence): This is a circular area around the target. If this area comprises a new sector, the track manager must register the target with the sector manager of this new sector.

The other element involved in the system, which carries out the simplest tasks, is the sensor agent. This agent carries out sensor measurements, prefiltering, eliminating information not related to the target's movement, saving power, and avoiding information overload. The sensor agent provides the track manager or the sector manager with tracking information. This kind of agent uses piggy-back technology to reduce the number of messages transmitted in the system by filtering routine messages.

The Directory Service provides agents with the ability to store textual information from many different sources and process requests for this information. It centralizes and spreads information, avoiding unnecessary interactions, and helps with agent self-discovery. It is important to emphasize that it does not store any type of identifier since the sensors lack their own identifier. Each agent has its small directory service where the sector manager descriptions are locally stored, providing mechanisms to find the sector manager and the neighbor sector managers.

The location and target speed are not directly predictable using the information provided by a sensor. To be able to know the target state, different models based on mathematical calculations are used:

- Process model: A general model of the object's path, in which it is possible to establish the location by means of the history of the target state.
- Time frames: A method based on the measurement delay obtained. With this information, the estimated location and speed of the object is updated.
- Location model: The measurements are stored in a queue every time frame.
- Amplitude handler: Amplitude measurements are used to determine the object's position.
- Frequency handler: Frequency values are used to determine the speed of the target since the frequency is a linear function of the radial speed.
- Motion model: A technique that learns from the movement of the object and is based on the supposition that the object is exposed to inertia laws and is not going to change its direction quickly.
- Target location: With a controller capable of receiving suitable measurements of different parameters, an object's position can be known.

The main problem with static agent applications is the resource assignment for multiple target tracking, which has the following two main difficulties:

- Sensor distribution, because the sensors may not be available and or there may not be sufficient nodes.

- An agent has partial knowledge of the problem but it needs to act using local information and information from neighboring agents.

The negotiation protocol proposed by Lesser et al. (2003) known as SPAM (Scalable Protocol for Anytime Multi-Level negotiation) has been used to manage a set of distributed sensors in order to solve the problem of multiple-target tracking. The main objective of the multiple-target tracking problem is distributing sensors to obtain the corresponding measurements for optimal tracking quality.

SPAM works in the following way: When a target is detected, an agent must be responsible for tracking it. To do so, it is necessary to determine the sensors that are in use for this task and to create a schedule of the available sensors. When there is more than one target, a conflict can arise in terms of sensor demand and assigning the available resources to the different elements involved in the tracking task. For this reason, track managers need to negotiate this allocation of resources in order to solve the problem in their local scheduler and/or to produce global scheduling in order to maximize tracking quality. The main idea of SPAM is to select one of the implied track managers as an intermediary to solve the problem and to provide partial solutions for the involved controllers. SPAM was designed under real-time running conditions. However, its complexity has led to the development of two new improved protocols: sequential SPAM and synchronous SPAM.

4.2.3.2 Applications with Mobile Agents

According to the proposal by Szumel et al. (2005), future networks will probably contain hundreds or millions of heterogeneous nodes. It is difficult to provide scalable solutions for energy- and bandwidth-limited WSNs. Therefore, agents can be used, since these programs can be present in any node, processing data and decreasing the interactions among nodes, as this activity consumes a great deal of energy.

The advantage of using agent-based solutions includes the support of multiple users, incremental programming, and efficient resource use. The agents only need the resources of the visited node such as its processing capacity and data transmission. New agents will be generated only in implied nodes when an object is detected. This makes successful target tracking possible. The major advantage can be seen when the network must be retasked, since using agents means that only some parts of the network need to be retasked. Szumel et al. (2005) have implemented this framework using Crossbow motes (MICA2DOT) running on TinyOS. The simulation has been carried out using TOSSIM, in a secure environment provided by a small variation of the Maté virtual machine.

Fok et al. (2005) describe Agilla, a mobile agent middleware that runs on TinyOS and provides an environment for mobile agent executions in WSNs. Agilla increases the flexibility of the network, simplifies the application development, and uses a virtual machine (such as Maté). The agent execution state is stored in couples, constituting a shared memory space inside the node. Agents may consult and obtain that shared memory asynchronously.

4.3 Design Strategies and Operation of WSN Software

4.3.1 Software Design Strategy in WSNs

A proposal for a possible WSN software design cycle can be found in Blumenthal et al. (2003). They proposed a software organization with an intermediation software layer (middleware) above the operating system. Its aim is to provide services to the applications. All the components of this architecture are represented by blocks. Figure 4.4 shows the proposed software development cycle for these types of networks.

The structure of the running software per Blumenthal et al.'s proposal is shown in Fig. 4.5.

With continuing advancements in sensor node design and increasingly complex applications, an interest in design automation of sensor network applications is inevitable. The objective is to eventually enable domain experts to be able to design and analyze algorithms, and automatically synthesize programs for an abstract machine model of the underlying system, without requiring knowledge of low-level networking aspects of the deployment.

4.3.2 Software Architecture in WSN

In typical WSN deployments, several types of applications and software coexist in different hardware platforms, such as sensor nodes, server nodes,

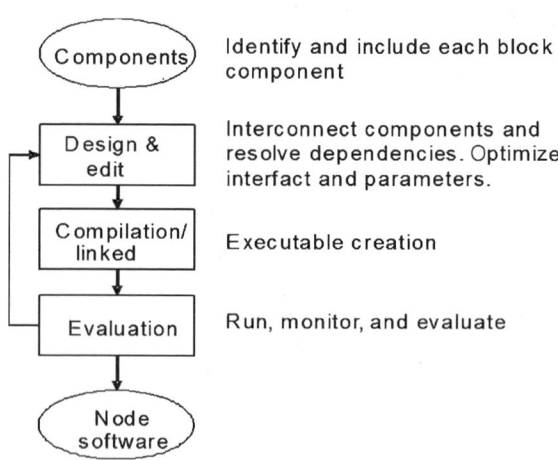

Fig. 4.4 Software design cycle for sensor nodes. (Adapted from [Blumenthal et al., 2003]. © 2003 IEEE.)

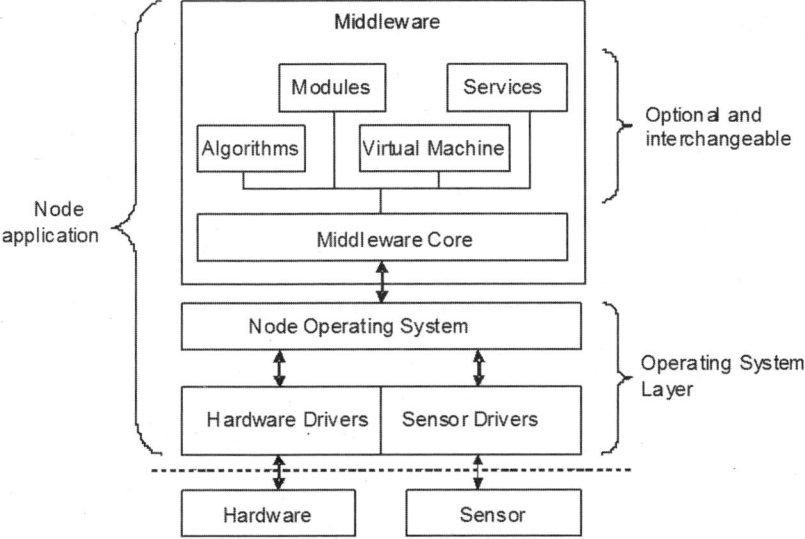

Fig. 4.5 Structure of an application running on a node. (Adapted from [Blumenthal et al., 2003]. ©2003 IEEE.)

or gateways, and client equipment. WSN architecture has different layers, as shown in Fig. 4.6. Several authors (Turon, 2005; Blumenthal et al., 2003) coincide with this layer decomposition, although there are some differences in nomenclature.

As depicted in Fig. 4.6, the software architecture consists of three different layers: the mote layer, the server layer, and the client layer.

- The mote layer is composed of the motes with their sensors. In this layer, the software needs to include the light operating system executed, e.g., TinyOS, and the corresponding applications necessary to obtain a service, i.e., environment monitoring or intruder tracking. These applications are usually developed for the specific hardware on which they are going to run, and the programming language used must fit well with highly restricted devices; nesC can be an adequate programming language. nesC (Gay et al., 2003) was created specifically to adapt application programming in embedded network systems, a category that includes WSNs. The main characteristics of nesC's design were inspired by the TinyOS operating system: event-based execution, incorporation of a concurrence model, and component-based application design. In fact, TinyOS has been reimplemented in the nesC language.

- The server layer receives the information from the WSN by using proprietary protocols and stores it in databases. It also offers services, usually by means of TCP/IP interfaces, in order to allow interested clients to access this information. In this case, the programming languages are selected not based on the

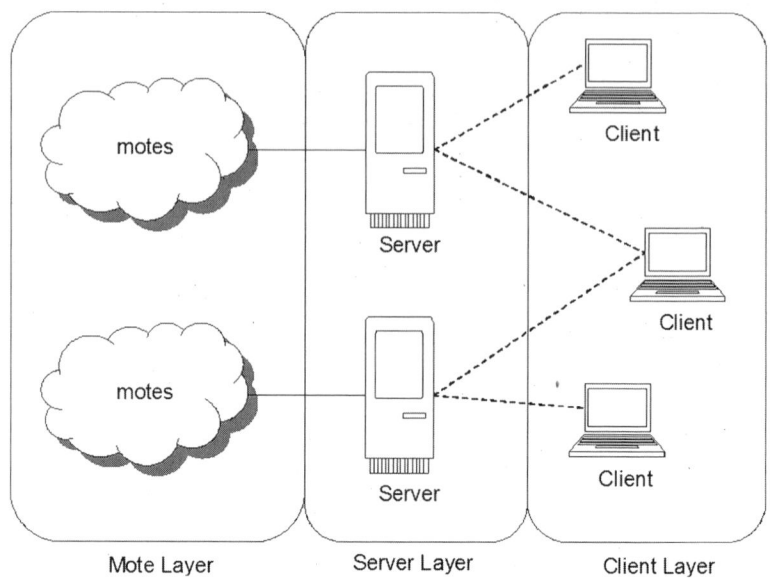

Fig. 4.6 Software architecture for a WSN. (Adapted from [Turon, 2005]. ©2005 IEEE.)

lack of resources, but rather on their portability to execute in machines with different characteristics and different general-purpose operating systems for different platforms.

• The client layer includes a graphical user interface that allows the information, topology, and state of the WSN as well as its management to be seen. This software must provide the user with the information needed for managing the WSN, interpreting the large amounts of information generated, and monitoring the network's health.

Brokerage Between Mote and Server Layers

The data that the sensor nodes, i.e., the motes, measure and transmit do not necessarily have to reach the server in their raw form, as seen in Fig. 4.6. In other words, there can be brokers, meaning service providers or intermediate processing entities, in charge of somehow filtering the motes' information and offering higher-layer services. Figure 4.6 shows how these brokers could be conceptually placed between the mote layer and the server layer, although physically these functions can be performed inside some of the sensor nodes.

One example of such an architecture including service providers inside the motes' network can be seen in the proposal from the EYES project (EYES, 2002), as shown in Fig. 4.7. In this architecture, both a "sensor and networking layer" and a "distributed services layer" are defined. The EYES project assumes heterogeneity in the nodes' capabilities, allowing the functions to be assigned to

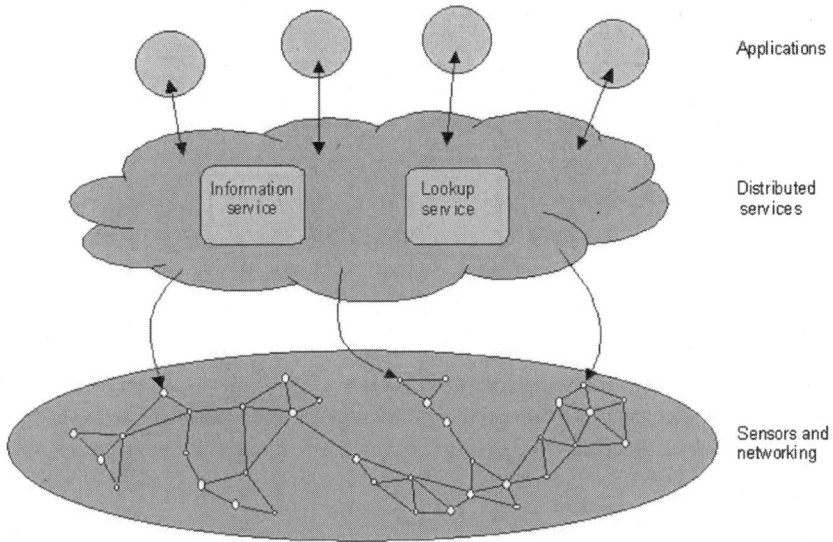

Fig. 4.7 EYES conceptual architecture. (From [EYES, 2002].)

the most adequate nodes. The nodes are arranged in clusters in order to offer the
needed abstract services. As an example of the considered services, the information
services may include information preprocessing, replication, or computations, thus
reducing the amount of data to be transmitted through the wireless channels.

For applications to access the services in a data-centric manner, the EYES
project also proposes using semantic addressing, in particular, using a content-
based publish-subscribe (CBPS) model (see (EYES, 2003)).

EYES is not the only project to propose the use of a publish-subscribe model.
The more recent e-SENSE project also includes this model (see (e-SENSE,
2006)), which the middleware subsystem of its architecture provides. The applications may access the data generated by the sensor nodes using the services
offered via the middleware. In this context, the broker appears as one of the
actors, sitting between the generators and the consumers of the information, and
is responsible for disseminating the publish events and the subscribe events to the
appropriate nodes. If the network is organized using a cluster topology, there
may be a broker for each cluster representing the nodes in that particular cluster.

4.3.2.1 Software in the Mote Layer

A division is possible in the mote layer among the possible software applications
that can be performed. On the one hand are sensing applications used to perform the final network functionality, such as tracking, monitoring, or any other
WSN application with the help of the intermediate software, i.e., the middleware, which provides services to those applications. We explained the state of

the art in middleware platforms for WSN in earlier sections, while we will not classify the specific application software because of its inherent application-dependent nature and variety. The specific application software will depend on the final WSN implementation goals and deployment.

On the other hand is the node's operating system (OS). The OS is the key to the performance of the distributed computing environment of wireless sensor networks. Some OS features, such as OS protection, virtual memory, preemptive scheduling, and others, will significantly improve the reliability of WSN systems and facilitate the development of complex WSN software. Several operating systems have been developed for sensor network applications:

- **TinyOS** (Szumel et al., 2005) is the most common operating system for WSNs today due to its open source and suitability for existing hardware platforms. It is widely used for simulating, developing, and testing algorithms and protocols. More than 500 groups of investigators and companies trust TinyOS and Berkeley/Crossbow's motes. TinyOS's architecture is based on components, and its design is specifically oriented to nodes with limited resources, such as the nodes of a WSN. The component libraries of TinyOS offer the following application network protocols: distribution services, sensor handlers, and data acquisition tools that can be used directly or be modified for a specific purpose. The model of execution by means of events allows precise, powerful, and flexible programming management, which is very necessary in WSNs due to their unpredictable nature, a feature from their real-world interface.
- **MagnetOS** (Barr et al., 2002) has a single-system image that creates the false impression of the Java virtual machine over a distributed sensor network, enabling programmers to use the Java language but supporting only limited heterogeneity and introducing considerable overhead on its instructions. The runtime system performs code partitioning and object placement through the network, hence reducing the energy expenditure. MagnetOS supports openness and scalability, providing dynamic run-time support on each node and service for application monitoring and object creation, invocation, and migration. A set of ad hoc protocols provide mobility and smart robust object migration strategies. The size of the additional overhead on its instructions makes this system inappropriate for mote-class sensor networks. Currently, a new native Java VM is under development.
- **SOS** (Han et al., 2005) is an operating system specifically designed for mote-class sensor networks. Furthermore, its kernel contains a sensor API that facilitates the development of WSN applications. SOS allows dynamic configurability, enabling programmers to change software modules in sensor nodes once they have deployed and initialized the network. SOS provides services such as incremental software updating, allowing the addition of new software modules and the removal of unnecessary ones. It implements an architecture composed of a common kernel and dynamic application modules that can be loaded and unloaded at run time.

- **t-kernel** (Gu and Stankovic, 2006) is a portable OS kernel used in WSN systems to perform extensive code modification at load time. The modified code and the operating system work together to support OS protection, virtual memory, and preemptive scheduling, three features that can considerably improve the consistency of WSN systems and make the development of complex WSN software easier. The t-kernel enables developers to design more consistent and complicated sensor networks. It also incorporates new design techniques, such as efficient binary translation on highly constrained sensor nodes, differentiated virtual memory without repeatedly writable swapping devices, and the protection of the OS from application errors without privileged execution hardware.
- **Contiki** is a portable operating system designed specifically for resource-limited devices such as sensor nodes. It is built around an event-driven kernel but supports preemptive multithreading on a per-process basis. It also supports a full TCP/IP stack. Contiki only requires a few kilobytes of code and a few hundred bytes of RAM.
- **eCos** (eCos) stands for "embedded configurable operating system." This is an open-source, royalty-free, real-time operating system intended for embedded systems and applications that need only one process. The OS is highly configurable and can be customized for precise application requirements with hundreds of options, delivering the best possible run-time performance and minimized hardware needs, which is a requirement for WSNs. eCos was designed for devices with a memory size of tens to hundreds of kilobytes or with real-time requirements.
- **Cormos** (Yannakopoulos and Bilas, 2005) is a run-time system for sensor nodes with processing and radio-frequency capabilities. The system provides easy-to-use abstractions that integrate communication with event processing. It is modular, uses a unified interface for system and application components, and is designed for systems with stringent resource limitations.
- **MantisOS** (Mantis) (multimodal system for networks of in situ wireless sensors) provides a new multithreaded cross-platform embedded operating system for wireless sensor networks. Sensor networks accommodate increasingly complex tasks such as compression, aggregation, and signal processing. Therefore, the preemptive multithreading in the Mantis sensor OS (MOS) enables microsensor nodes to natively interleave complex tasks with time-sensitive tasks, thereby reducing the bounded buffer producer-consumer problem. To achieve memory efficiency, MOS is implemented in a lightweight RAM footprint that fits in less than 500 bytes of memory—this includes the kernel, scheduler, and network stack. In order to improve energy efficiency, the MOS power-efficient scheduler determines if it is safe to put the microcontroller to sleep with the MOS sleep() function, after gathering information from all active threads and reducing current consumption to the μA range. A key MOS design feature is flexibility in the form of cross-platform support and testing across PCs, PDAs, and different microsensor platforms. Another key MOS design feature is support for the remote management of in situ sensors through dynamic reprogramming and remote log-in.

- Other operating systems for WSNs include **Bertha**, which is a pushpin computing platform, **BTnut Nut/OS, EYESOS, SenOS,** and **LiteOS**.

Some of the environments and high-level languages have already been described, such as Maté, TinyDB, and the COUGAR project proposal; other examples are summarized below.

TOSSIM

(Levis and Lee, 2003). This proposal is a discrete-event simulator that allows applications programmed for TinyOS to be evaluated before deploying them in a real network. Applications can be debugged by means of this software executed on a PC platform, compiling them for TinyOS. The emphasis has been put on a precise simulation of the TinyOS operating system, whereas the simulation of the external effects of the environment such as battery consumption, propagation models, etc. is less precise. For this reason, TOSSIM is not recommended as the only way to evaluate and debug an application.

SeNeTs

(Blumenthal et al., 2005). An environment that allows WSN applications to be evaluated and debugged, SeNeTs can be executed on diverse distributed platforms, which increases its scalability. Other advantages that Blumenthal et al. (2005) have highlighted are the hardware platform independence of the sensor nodes, whereas TOSSIM is very tied to TinyOS, and the motes' concrete hardware.

WSN application evaluation with SeNeTs is done by grating the application software running in a sensor node with an adaptation software with the following components (Blumenthal et al., 2005):

- Hardware abstraction such as sensor, memory, communication devices, etc.
- Environment emulation: Unlike TOSSIM, SeNeTs includes the environmental representation as an evaluation platform requirement. This emulation module allows more valid results to be obtained than those obtained when the real environment of the application execution is not taken into account.
- Other components: tracing, control, debugging, etc.

EnviSense

In the same work (Blumenthal et al., 2005), the authors indicate that they are developing EnviSense, which is a graphical environment unlike the SeNeTs' textual interface, in order to configure, visualize, and obtain data from WSNs. Nevertheless, EnviSense does not provide mechanisms to modify the applications.

4.3.2.2 Software in the Server and Client Layers

As stated above, server-layer software applications offer services, usually by means of TCP/IP interfaces, allowing clients to access the collected WSN information. The client layer includes a graphical user interface that visualizes the information, topology and state of the WSN as well as its management.

The following paragraphs summarize the characteristics of several WSN deployment and operation tools.

TASK

TASK (Tiny Application Sensor Kit; see (Buonadonna et al., 2005)) is a toolbox developed by Intel Research at Berkeley whose goal is programming and deploying WSN applications without having low-level knowledge of these types of systems. The components of this architecture include

- WSN based on TinyDB that allows programs to access a pseudo-SQL interface.
- TASK server placed in the gateway, which connects the WSN to the Internet and allows external access to measurement data using standard interfaces, i.e., HTTP.
- TASK DBMS, a relational database that stores information about the sensor measurements, their location, etc.
- TASK client tools incorporate various tools for deployment, configuration, and results visualization.
- TASK Field Tool, executed on a PDA-like device, allows problems in certain areas of the network to be diagnosed and solved.

Mote-View

Crossbow's (Crossbow, 2007) graphical tool to facilitate WSN management and operation, MOTE-VIEW is client-layer software (using the nomenclature in Fig. 4.6) and is structured in four extensible modular layers by means of adding plug-ins:

- Data access abstraction layer: Across this layer other MOTE-VIEW modules do not need to know the internal information structure offered by different servers similar to those in Fig. 4.6.
- Node abstraction layer: Across this layer it is possible to access and configure the state of the nodes. By adding plug-ins, the node abstraction layer can support nodes from different manufacturers.
- Calibration and units conversion layer: This layer uses calibration coefficients depending on the nodes and applies them to the raw data obtained from the data access abstraction layer in order to show the measurements in engineering units.

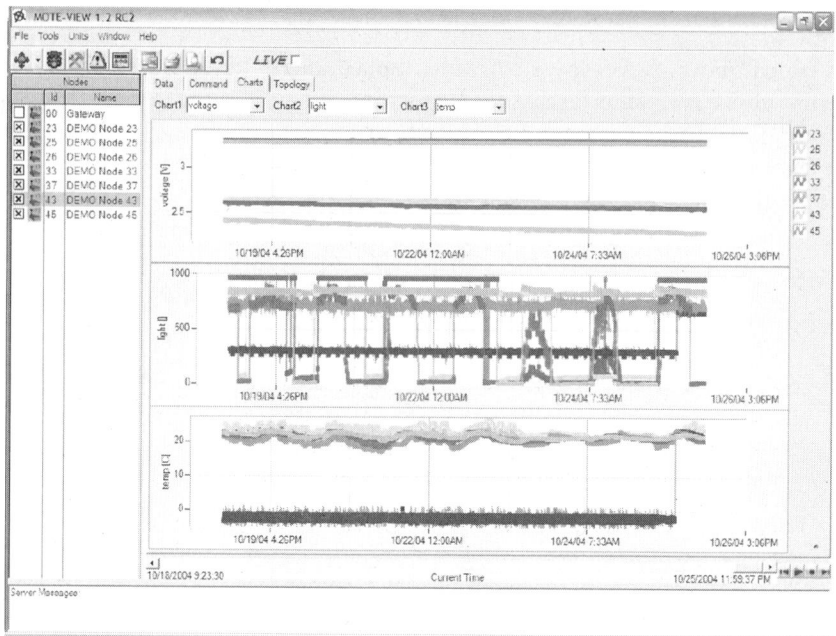

Fig. 4.8 Chart view of MOTE-VIEW. (From "MOTE-VIEW 1.2 User's Manual. Revision B." January 2006. ©Crossbow Technology, Inc.)

- Visualization abstraction layer: This layer offers three different graphical data representations:

  – Data: recent readings of the network sensors by means of a spreadsheet representation
  – Chart: representation of the measurements of a specific sensor in a certain time interval (see Fig. 4.8)
  – Topology: visualization of the WSN topology

This software can also provide indications about the sensors' health. MOTE-VIEW includes various tools to configure a mote (with "MoteConfig") by downloading precompiled applications for TinyOS and loading them into a node or saving the received measurements from a gateway in a database or file (using the "XServe" utility).

Stargate

Stargate NetBridge (Stargate) is a specific platform based on an Intel processor and produced by Crossbow on which a Linux operating system is executed. This platform can act as server in a WSN based on TinyOS. The platform also provides two software applications that make the data directly accessible by a

Fig. 4.9 Stargate server with a mote acting as base station. (From [Crossbow, 2006]. ©Crossbow Technology, Inc.)

client using a Web interface. These applications are called XServe (management tool) and MoteExplorer (data visualization tool).

Stargate NetBridge is actually the new version of a previous product called Stargate Gateway, in which the received mote data are saved in a Postgres database by a software tool named Xlisten.

To obtain the WSN data, it is possible to connect a mote acting as base station to a Stargate board (see Fig. 4.9). In addition, to offer this data to the clients executing **MOTE-VIEW**, for example, Ethernet, WiFi or cellular telephony can be included in the platform's standard interfaces.

Janus

One proposal for monitoring and configuring the sensor network is Janus (Gold, 2005), which is an architecture that provides flexible access to sensor networks. The existing approaches for remote access to sensor networks are typically application-specific. However, Janus tries to provide a flexible signaling mechanism that supports both passive and active access approaches, uses a generic RPC-like mechanism, and enables on-demand setup of access to sensor network resources of varying types.

(Balani et al., 2006)

Balani et al. proposed a multilevel software reconfiguration for sensor networks. This system supports software reconfiguration in embedded sensor networks at multiple levels and is capable of reprogramming or reconfiguring the sensor network after deployment. The system architecture is based on an operating system consisting of a fixed, tiny, static kernel and binary modules that can be dynamically inserted, updated, or removed. On top of the operating system is a command interpreter, implemented as a dynamically extensible virtual machine, that can execute high-level scripts written in portable byte code.

4.4 WSN Simulation Platforms

4.4.1 Importance and Challenges of WSN Simulators

Wireless sensor networks have tremendous potential to monitor, study, and analyze phenomena in the physical world in detail never before available, in places too far, too deep, too high, or too dangerous for researchers to go. Simulation can be of great help to ensure the shortest possible time to market and to minimize the overall cost of WSN design. Being that the cost, time, and complexity involved in deploying and constantly changing large-scale WSNs are prohibitively high, simulation is a cost-effective choice for the rapid exploration and validation of WSN applications. Simulation provides controlled and repeatable environmental conditions for evaluating and optimizing the design parameters and/or the configuration alternatives. It also offers very good insight into the effects of the various parameters and thus helps identify those that have the greatest importance for the system's operation.

The simulation of wireless networks is inherently different from that of wired networks. The signal interference and attenuation concerns are more complicated for wireless media than for wired media. The broadcast nature of wireless radio transmission also makes communication topology in simulation models relatively denser than for an equivalent wired network. The computation effort of WSN simulators is dramatically increased by specific features like node mobility, distributed behavior, power, and terrain models. Consequently, accurate fine-grained WSN simulations present a significant challenge.

A wide range of simulation tools specially adapted to WSNs has already been presented in the literature. A high-level description of a possible WSN simulation framework can be found in Park et al. (2001) and Sobeih et al. (2006). The sensor nodes detect the stimuli, i.e., signals, generated by the target nodes over a sensor channel and forward the detected information to the sink nodes over a wireless channel. Two different models for signal propagation are therefore included: a sensor propagation model and a wireless propagation model.

The existing simulator tools are either commercial or open source and are mainly developed in Java or C++. These simulators strongly differ with respect to features such as

- Scalability: The capacity to simulate a high number of nodes with sufficient precision in a finite time.
- Real-time requirement: Real-time or close to real-time simulation can be related to the simulation architecture and the programming language and to some other features as well.
- Software emulation: The possibility to run the various software (OS, middleware, etc.) during simulation as if they were running on the real processors—also called "software in the loop."
- Hardware emulation: The ability to accurately simulate the behavior of various hardware parts of the node and to evaluate some important parameters such as consumption, memory use, collisions, etc.
- Model fidelity: Availability of detailed models for various aspects of networking such as propagation, protocols, mobility, etc.
- Reliability study: Possibility to simulate the occurrence of various faults or defects, i.e., hardware, physical, noise, etc., during the network simulation and to visualize their effects on the network's behavior or on some parameters such as consumption, delays, etc.
- X in the loop: The possibility of connecting other (hardware and/or software) devices to the simulator for various purposes; simulation-time reduction by replacing a hardware simulation model by a real hardware system, either existing or simulated on an FPGA or other hardware platform, performance analysis of protocols by injection of real signals, reliability study in the presence of real perturbations, etc.

Ahead we give an overview of the most relevant WSN simulators and describe their advantages and drawbacks. We characterize the simulators and identify those most suited for research.

4.4.2 Review of WSN Simulators

Most sensor networks simulators are discrete-event simulators. In discrete-event simulation, the operation of a system is represented as a chronological sequence of events. Each event occurs at a point in time and marks a change of state in the system. A discrete-event simulator proceeds by constantly removing the current event from the head of its time-ordered event queue and then simulating that event. The model is then advanced to the time of the next significant event. Hence, if nothing is going to happen for the next three time units, the simulation kernel will move the model forward three time units at once. The nature of the jumping between significant points in time means that in most cases the discrete-event mechanism is more efficient and allows models to be evaluated more quickly. A good description of discrete-event simulation can be found in Schriber and Brunner (2006).

There are a few commercial sensor network simulators such as OPNET and QualNet. The other simulators are open-source codes or academic simulators. Commercial products are very powerful and complete systems that are able to simulate a large number of nodes, up to several thousand, in real-time constraints. They offer high portability, i.e., run on Linux, Solaris, Windows, and Mac OS, have a great number of model libraries such as propagation, terrain, protocols, etc., and offer a variety of software components for WSN developers such as scenario designer, graphing tool and data analyzer, communication tracer, etc.

4.4.2.1 Commercial Simulators

The use of commercial simulators may not be suited for a research project. For simple cost reasons as well as openness and evolution capacity reasons, the use of easily upgradable simulation software is preferable. Nevertheless, commercial products are supposed to be more mature products that can be applied to a wide range of problems. This is why it can be very interesting to have a precise knowledge of their capacities and the way they run the simulations.

OPNET

(OPNET) is a commercial network modeler and simulator provided by OPNET Technologies, Inc. OPNET models a network in a hierarchical approach that closely matches the hierarchical architecture of the Internet: networks, subnets, and nodes (fixed, mobile, or satellite). Each node is modeled as a set of processes where each process is modeled as a finite state machine (FSM). The entire network is simulated using a discrete-event simulator. Nodes in conventional OPNET models are connected by static links. OPNET supports three types of links: point-to-point, bus, and wireless. A wireless link is used in wireless, mobile, or satellite network simulation. OPNET uses a 13-stage "transceiver pipeline" to dynamically determine the connectivity and propagation effects among nodes. Users can specify the transceiver frequency, bandwidth, power, and other characteristics. These characteristics are used by the transceiver pipeline stages to calculate the average power level of the received signals to determine whether the receiver can receive this signal. In addition, antenna gain patterns, bit errors, and terrain models are well supported. However, although OPNET can simulate both wired and wireless networks, it does not include detailed models for WSNs.

QualNet

(QualNet). Provided by Scalable Networks Technologies, Inc., QualNet is network modeling software that predicts the performance of networks through simulation and emulation. It enables real-time network simulation, which supports hardware simulation, software simulation, and human-in-the-loop simulation for networks of up to several thousand nodes. A key functionality is extensibility, or the ability to interface to other simulations and real networks, which greatly increases the value of communication simulations.

4.4.2.2 Open-Source or Academic Simulators

Some of the most relevant academic WSN simulators are presented below. Often these simulators are still under development; some of them are well suited to help with research.

OMNeT++

(Varga, 2001). OMNeT++ is an open-source tool that shares many concepts, solutions, and features with OPNET. OMNeT++ is a discrete-event, component-based, modular, and open-architecture simulation environment with strong GUI support and an embeddable simulation kernel. OMNeT++ provides component architecture for models. Components, i.e., modules, are programmed in C++ and then assembled into larger components and models using a high-level language (NED).

GloMoSim

(Global Mobile Information System Simulator) (GloMoSim, 2001; Bajaj et al., 1999). A simulation environment for purely wireless mobile networks, GloMoSim was designed as a set of modules in an architecture structured into eight layers. Each module simulates a specific protocol in the protocol stack. GloMoSim has been designed using the parallel discrete-event simulation capability provided by PARSEC (PARSEC), a C-based sequential and parallel simulation language that can be used to program new modules that can be added to GloMoSim. GloMoSim offers different protocols to model node mobility and radio communication. GloMoSim has already been used to simulate networks with thousands of wireless nodes and provides a rich set of models for both existing and novel protocols at multiple layers of the protocol stack. Apparently, GloMoSim does not offer environment or power models.

Ptolemy

(Ptolemy). This is an ongoing project at UC Berkeley that studies the modeling, discrete-event simulation, and design of concurrent real-time embedded systems. The key underlying principle in Ptolemy is the ability to use multiple computation models (e.g., continuous-time, data flow, finite state machine) in a hierarchical heterogeneous design environment. Ptolemy does not support network emulation but does support both wireless network and sensor network simulations.

ns-2

(ns-2). This is a discrete-event simulator that provides support for TCP, routing, and multicast protocols, among many others. The support for wireless and mobile network simulation provides various modules for mobile wireless network simulation, such as radio propagation models, the IEEE 802.11 MAC protocol, mobility models, different ad hoc routing protocols (e.g., AODV and DSR), and Mobile IP. The latest version of ns-2 supports the simulation of pure wireless LANs, multiple-hop ad hoc networks, and combined simulation of wired and wireless (known as "wired-cum-wireless") networks. Maintaining real code in ns-2 is not transparent.

VisualSense

(Baldwin et al., 2004). VisualSense is a modeling and simulation framework that builds on and leverages Ptolemy to support design, simulation, and visualization of sensor networks. A sensor node can be modeled either in Java or by using conventional discrete-event models (e.g., block diagrams) or Ptolemy models (e.g., continuous-time or data-flow models). VisualSense supports sensor and wireless channels, antenna gains, terrain models, and battery models.

SensorSim

(SensorSim; Park et al., 2001). From the University of California at Los Angeles (UCLA), SensorSim has further extended ns-2 by including support for sensor network simulation. It includes the definition of target, sensor, and sink nodes, sensor and wireless communication channels, physical media, a mobility model, and a power model. SensorSim has a more developed sensor model and power modeling capabilities. It is heavily integrated with the Sensor-Ware middleware layer and does not integrate easily to simulate systems running TinyOS or EmStar. Unfortunately, SensorSim has not been publicly released and does not appear to be under further development.

SWAN

(Liu et al., 2001). SWAN is a scalable sensor network ad hoc simulator that uses handcrafted models of the nodes. The simulator is built by putting together a high-performance scalable simulator from Dartmouth College and a portable router from BBN Technologies. The simulator provides the infrastructure for data exchange and for the synchronization of all components. The router is portable across different wireless platforms and also easily transportable into simulation testbeds, allowing the direct execution of routing algorithms at the source code level to take place. SWAN could handle networks with 10,000 nodes.

SWANS

(SWANS). A scalable WSN simulator built on the JiST (Java in Simulation Time) platform, SWANS is organized as independent software components that can be arranged to form complete wireless network or sensor network configurations. SWANS implements a data structure called "hierarchical binning" for efficient computation of signal propagation. SWANS can simulate networks that are one or two orders of magnitude larger than what is possible with GloMoSim and ns-2 using the same amount of time and memory with the same level of detail. SWANS could handle networks with 50,000 nodes.

SENS

(Sundresh et al., 2004; SENS). A platform-independent WSN simulator that uses handcrafted C++ models of the sensor nodes and emphasizes extensibility, reusability, and scalability. A component-port model makes simulation models extensible, where a new component can replace an old one if they have compatible interfaces and inheritance is not required. A simulation component classification makes simulation engines extensible in which advanced users have the freedom to develop new simulation engines that meet their needs. The component-port model frees the simulation models from interdependency and promotes reusability in this way. The modeling-based approach allows the evaluation of network delays, throughputs, packet collisions, and node localization errors for 10,000 nodes.

SENS has a modular, layered architecture with customizable components that model an application, network communication, and physical environment. By choosing appropriate component implementations, users may capture a variety of application-specific scenarios with accuracy and efficiency tuned on a per-node basis. To enable realistic simulations, values from real sensors should be used to represent the behavior of component implementations. Such behavior includes sound and radio signal strength characteristics and power usage. SENS defines the environment as a grid of interchangeable tiles to modulate sound and radio propagation at different levels of detail. Tile implementations for concrete, grass, and walls are available. Users may define other tiles to suit their needs.

J-Sim or Java-Sim

(Sobeih et al., 2006; J-Sim). An open-source component-based framework for WSN simulation and emulation, J-Sim has been developed entirely in Java. Coupled with the autonomous component architecture, this makes J-Sim a truly platform-neutral, extensible, and reusable environment. The framework provides an object-oriented definition of (1) target, sensor, and sink nodes, (2) sensor and wireless communication channels, and (3) physical media such as seismic channels, mobility models, and power models, i.e., both energy-producing and energy-consuming components. Customized application-specific models can be readily defined and implemented by subclassing appropriate classes in the simulation framework and customizing their behaviors.

EmSim

(Girod et al., 2007). EmSim provides a real code simulation capability for microserver applications implemented within the EmStar/Linux framework. Systems in EmStar are composed of many Linux processes, generally one process per separable service or module. EmStar services can be written in any language as long as they communicate using the EmStar IPC mechanisms. EmSim has many features in common with SensorSim, including the ability to run hybrid simulations with real nodes alongside simulated nodes. EmSim cannot assume as much about the uniformity of the application, because EmStar code can be written in a variety of languages. Thus, rather than compiling everything into one process, EmSim runs multiple simulated nodes as separate process trees. Each simulated node communicates with other simulated nodes through standard EmStar IPC channels. This multiprocess approach entails some scaling limitations. The lowest layer of services is provided by simulation components that provide separate device interfaces for each simulated node. For instance, the RF channel model module provides a packet-level "Link Device" interface for each simulated node and transfers packets among the simulated nodes according to a channel model. EmSim currently includes simulation components to support sensors and the RF channel. The RF channel simulator can assume one of several "personalities," including the standard TinyOS MAC (Levis et al., 2003), S-MAC (Ye et al., 2002), and 802.11 and supports a variety of propagation models tuned for different hardware types. EmSim does not attempt to model the MAC layer in detail, apart from aspects such as CSMA vs. TDMA and collisions. One of the key features of EmSim is its ability to support the emulation mode, where centrally simulated nodes interact using real radio or sensor hardware embedded in a target environment. The emulation mode is made possible in part by the fact that EmSim natively runs in real time rather than according to a virtual clock. The advantage of running in real time is that it enables interaction with external hardware or whole systems that also run in real time. The primary disadvantage is scalability. An experimental addition to EmStar called TimeWarp addresses these problems.

TOSSIM and ATEMU

(Levis and Lee, 2003; atemu). These are simulators for wireless sensor networks in which the nodes are Crossbow AVR/MICA2 motes, which will be simply referred as "motes." Motes have limited resources; at the core is an 8-bit 7.3728-MHz AVR microcontroller with 4 kB of main memory for stack and heap, 128 kB of program storage for code and pre-initialized data, 4 kB of nonvolatile EEPROM storage, and internal devices such as clocks and a serial port for controlling external devices. Software for the motes is generally built with either TinyOS, a set of components for building sensor network programs written in the NesC (Gay et al., 2003) programming language, or SOS (SOS), a light-weight, modular operating system designed for dynamic flexibility.

TOSSIM (Levis and Lee, 2003) is a discrete-event real code simulator. It provides a high degree of accuracy by using models with only a few low-level components and otherwise running the application source code unchanged. Specifically, TOSSIM compiles nesC source code together with TinyOS libraries into a binary code for the development workstation, replacing the software modules that create an interface hardware with emulation libraries, including timers, communication channels, sensors, and radio. TOSSIM preserves details of the MAC layer in the simulation, because the TinyOS MAC layer itself is part of the simulated application. TOSSIM models the wireless network with a directed graph, where each vertex is a node and each edge has a bit error probability in order to model radio transmissions. TOSSIM's level of detail was sufficient to measure packet losses, packet CRC failure rates, and the length of the send queue for up to 8,192 nodes (Levis and Lee, 2003). However, TOSSIM's compilation step loses the fine-grained timing and interrupt properties of the code that can be important when the application runs on the hardware and interacts with other nodes.

It is easy to go back and forth between a set of real motes and a TOSSIM simulation because the same NesC application code runs on real motes and in the TOSSIM environment. TOSSIM also has the capability to inject traffic into a simulated mote network in order to simulate tasking and other stimuli from external sources. However, it is not always convenient to feed TOSSIM with a task load generated by a complex external system.

Tython (Demmer and Levis, 2004) is an extension to TOSSIM that enables easy scripting of packet injection into the simulated mote network. However, encoding the complete behavior of the outside world into Tython may prove difficult.

ATEMU (atemu) is a hardware-level simulator for motes. It emulates the operation of the various components on a sensor node, such as the processor, timers, and radio interface. These emulations of individual sensor nodes are then tied together via their interactions with each other to form an emulation of an entire sensor network. In order to achieve this goal, ATEMU provides an extensible model of the air to simulate the wireless medium's operation. ATEMU can simulate sensor network programs with accuracy down to the clock cycle of each individual node. ATEMU's fine-grained accuracy enables a reliable count of the number of backoffs after transmission collisions for up to 120 nodes (Polley et al., 2004). ATEMU is particularly useful in studies monitoring power consumption or those involving heterogeneous.nodes in the same network, in terms of both software and hardware. This is not the case with TOSSIM, which is limited by being able to run only a single binary image on all sensor nodes and therefore not for use to perform studies where an alternate application is being used. However, ATEMU is 30 times slower than TOSSIM.

Included in the ATEMU distribution is the XATDB graphical front-end, which provides a complete system for debugging and monitoring the execution of the code. The sensor node binaries are executed on the ATEMU emulator.

Avrora

(Avrora; Titzer et al., 2005). Avrora is a sensor network simulator that is cycle-accurate like ATEMU and scalable like TOSSIM. The event-queue model for cycle-accurate simulation of device and communication behavior allows improved interpreter performance and enables an essential sleep optimization. Avrora allows date-/time-dependent properties of large-scale networks to be validated. A highly accurate energy model is available, enabling power profiling and lifetime prediction of sensor networks. Distance attenuation for multiple-hop scenarios is also modeled. However, Avrora does not yet model node mobility. Another phenomenon that is not modeled is clock drift, which takes into account that nodes may run at slightly different clock frequencies over time due to manufacturing tolerances, temperature, and battery performance.

The scale and run-time performance of wireless network simulators can be improved using a technique called staged simulation (Walsh and Sirer, 2004). The central idea behind staging is to eliminate redundant expensive computations through function caching and reuse. Staging is a general technique that retains the original accuracy of a nonoptimized simulator and is applicable to a wide range of simulators, including parallel and distributed simulation engines.

SNS

(SNS). A freely available staged simulator based on ns-2, SNS improves the total run time of the standard ns-2 simulator from $O(N^2)$ to $O(N)$. Consequently, SNS executes $\sim$50 times faster than regular ns-2 on a specific ad hoc network simulation setup with 1,500 nodes. This level of performance allows better scalability and translates into 10,000 nodes being simulated in less than one hour in SNS.

4.4.3 Conclusions on the Use of WSN Simulators for Research

In the context of research, some features of the WSN simulators presented previously are more important than others. For instance, whenever the simulator is to be used as the hardware abstraction layer (HAL) and as a design tool, as well as for a reliability study, software and hardware emulation would be critical. The other characteristics are less important for this specific example.

As far as using the simulator as a design tool, two options are possible: Build a new simulator from scratch, or upgrade an already existing simulator—one of the open-source simulators presented above. This option will be discussed and decided later. Some features of the open-source simulators presented above that could be used in the context of a research project are summarized in Tables 4.1 and 4.2.

Table 4.1 Main Features of Some Open-Source WSN Simulators That Could Be Suitable for Research Studies—Part 1

Simulator	OMNET++	GloMoSim	VisualSense	SWANS	SENS
Source reference	(OMNeT++)	(GloMoSim, 2001)	(Ptolemy)	(SWANS)	(SENS)
Year	1999	1999	2004	2004	2004
Development language	C++	C	Java	Java	C++
Node models	Flexible component-based architecture	Discrete-event model	Flexible	Discrete-event model	Flexible
Protocols	TCP/IP, SCSI, FDDI, Ethernet, Token Ring, GSM	Flexible	Flexible, heterogeneous	TCP, DSR, AODV	Flexible
Software emulation	–	–	No	–	–
Hardware emulation	–	–	No	–	–
Mobility models	Yes	Yes	Dynamically changing inter-connection topologies	Yes	–
Power models	–	–	Yes	–	Yes
Battery models	–	–	Yes	–	No
Radio models	–	Yes	Yes	Yes	Yes
Environment (terrain) models	–	–	Yes	–	Yes
Scalability (number of nodes)	–	~1,000	Only examples with few nodes	~50,000	~10,000

The symbol "–" indicates that information is not available.

Table 4.2 Main Features of Some Open-Source WSN Simulators That Could Be Suitable for Research Studies—Part 2

Simulator	J-Sim	ATEMU	Avrora	SNS
Source reference	(J-Sim)	[atemu]	(Avrora)	(SNS)
Year	2006	2004	2005	2003
Development language	Java	C	Java	C++
Node models	Flexible	(Extendable) Mote*-based	Mote*-based	Discrete-event model
Protocols	Flexible (IEEE 802.11 MAC, AODV,GPSR)	Various networking protocols (supported by TinyOS)	Protocols supported by TinyOS	Like Ns-2: IEEE 802.11 MAC, TCP, mobile IP, ad hoc routing, multicast
Software emulation	Yes	Instruction-level	Instruction-level	Interaction with a live network
Hardware emulation	–	Cycle-accurate	Cycle-accurate	–
Mobility models	Yes	–	No	Yes
Power models	Yes	Yes	Yes	–
Battery models	Yes	–	–	–
Radio models	Yes	Yes	Yes	Yes
Environment (terrain) models	Yes	Extensible model of the air	Yes	–
Scalability (number of nodes)	~100	~100	~10,000	~10,000

The symbol "–" indicates that the information is not available.
* indicates Crossbow AVR/MICA2 motes.

4.5 Open Issues in Software Technologies

4.5.1 Software Design and Development for WSNs

The commercial success of WSN applications depends on the availability of tools and methodologies for developing software, abstracting low-level details, and deploying and operating the network comfortably. Examples of this for PCs instead of WSNs can be Eclipse (see www.eclipse.org), Visual Studio (see www.microsoft.com), and other development platforms that provide a range of tools offering various benefits for developers. For example, if we had a development/programming environment for WSNs in which the syntax was corrected, it could compile a program and test it in a virtual machine that could simulate a single mote or a set of interconnected ones. This in turn would allow the above-mentioned application to be loaded into a mote, the development of software to be simpler, and the commercial success of WSN to be easier. Nowadays, since WSNs continue to be an active research area, there are no fully consolidated tools or toolboxes to aid software development for WSNs. Thus, it would be interesting to have an open development platform comprised of extensible frameworks, tools, the ability to test virtual machines, and run times for building, deploying, and managing software for WSNs.

Although there are some proposals regarding software design for WSNs, this continues to be an area in which interesting contributions can be made, especially regarding the generalization and definition of methodologies for generating reliable software. A proposal in this direction is the aforementioned Blumenthal et al. (2003) paper that proposed a possible design cycle for WSN software development.

4.5.2 Low-Level Detail Abstraction

There are essentially two types of proposals for abstracting the low-level details: virtual machines (VM) and intermediation software layers (middleware). Although there are proposals for both types of abstraction mechanisms in WSNs and they contribute many good ideas, none is a complete solution.

One of the best proposals among virtual machines is Squawk (Simon et al., 2006). Its main drawback is its limited field of applicability, since Squawk VMs are mainly applied to the Sun Small Programmable Object Technology (SPOT) wireless device (SunSpot), a device developed at Sun Microsystems Laboratories for experimentation with wireless sensor and actuator applications. Another example of the VM abstraction applied to WSN is Maté, also described earlier.

Some of the many proposals for WSNs in terms of middleware have not been implemented or tested in a real environment. The variety of the proposals can make selecting the most suitable one for a specific application difficult. We have described some examples of WSN middleware proposals: TinyOS (which includes some modules that could be regarded as belonging to a middleware layer), Impala, COUGAR, MiLAN (Middleware Linking Applications and Networks), SINA, AUTOSEC, DSWARE, Smart Messages Project, etc.

Wolenetz et al. (2005) show a simulation-based study of some WSN middleware proposals, especially DFuse, IrisNET, and SensorWare. The authors have simulated a surveillance application workload with middleware capabilities for data fusion, adaptive policy-driven migration of data fusion computation across network nodes, and pre-fetching of streaming data inputs for fusion processing. Their study sheds light on important application features such as latency, throughput, and lifetime with respect to migration policy, node CPU, and radio, but only includes a limited amount of middleware proposals.

A paradigm that has been proposed many times for WSNs is the "publish-subscribe paradigm." Two main contributions in this direction can be mentioned:

- MIRES (Souto et al., 2005). Some of MIRES's shortcomings are the following: It is implemented on top of TinyOS, needs improvements to make it more robust to sudden topology changes and individual node crashes, and has not yet been tested using real sensor nodes.
- SensorBus (Ribeiro et al., 2005). This seems a good proposal but is still under development.

The "publish-subscribe" paradigm is very interesting for consideration in a WSN since it is closely associated to an event-based model appropriate for many typical WSN applications. By using this model, communications are performed in an asynchronous manner, effectively decoupling the publishers of "services" from the "subscribers." One advantage of this approach is an enhanced robustness. The service will be available as long as there is at least one entity or group of entities that publish its availability, and no changes have to be made to the subscribers if a different set of entities offer the same service.

4.5.3 Software Deployment and Operation in WSNs

Software deployment and operation in WSN is not a frequent area of research. The main proposals of deployment and operation focus on the physical deployment of the nodes, and not on the software deployment inside the motes.

A proposal for monitoring and configuring sensor networks is Janus (Gold, 2005). The Janus approach is lightweight, flexible, and extensible, but only its ability to support client interaction with a prototype sensor network implementation, as opposed to using an actual sensor network, has been demonstrated. Nevertheless, this is an interesting architecture for the deployment and operation of WSNs. Another interesting proposal in the deployment and operation areas is Balani et al. (2006), a multilevel software reconfiguration for sensor networks.

Some proposals to simulate the network could help with efficient deployment issues. For instance, MSSN (Middleware Simulator for Sensor Networks), is an event-driven middleware simulator proposed in Wolenetz et al. (2005) with similarities with other WSN simulators: GlomoSim, NESLsim, ns2-wireless, TOSSIM, and Prowler (Wolenetz et al., 2005). All of these simulators have two main drawbacks:

- None covers all possibilities; for example, some of them exclusively support a single manufacturer or only a subset of the possible topologies, or they only simulate TinyOS.
- They simulate the link and network levels but do not pay enough attention to the application layer.

In some networks, an over-the-air programming of the nodes could be desirable, being able to change the nodes, and thus the network behavior, dynamically. An important proposal, which can also be addressed as a type of middleware, is the use of mobile agents capable of moving between the nodes and actually changing the nodes' behavior.

In fact, the use of the multi-agent paradigm (whether mobile or not) in wireless sensor networks is an interesting possibility that still has open research issues. For instance: How complex is the processing carried out by the motes' agents compared with the processing at the server? More complex processing in the motes can reduce the used bandwidth but increases the processing energy. How does an application programmer use the agent paradigm to design the application? Are the agents mobile or fixed? Is an agent always active or is its activity triggered by certain events, for example, by the presence of a target that must be tracked? Can the agents be dynamically deployed in the nodes or not? Is the same set of agents present in every node?

4.5.4 Quality of Service (QoS)

To provide QoS to a WSN, the lower levels of the architecture, which are mainly the link and network layers, have to be designed to take into account the traffic requirements. However, these are not the only layers in which the QoS issues have to be considered. For some applications, there has to be a way of indicating the degree of requested QoS. These requests should be formulated using a high-level abstraction and not by burdening the application programmer with the details of the lower layers. Thus, a first issue that arises is how to represent the QoS capabilities that an application may access in a WSN in a general but useful manner. This high-level representation of the QoS parameters also has to be mapped to lower layers and network-related representations of QoS. In other words, the middleware has to offer the necessary abstraction of the QoS services to the application, thus effectively hiding the low-level details (a first such approach—that has yet to be implemented—is Sharifi et al. (2006); it is

valid for cluster-based networks). Undoubtedly, the QoS aspects in any network with such strong resource restrictions as a WSN are still a major open
research issue.

4.5.5 Application Software

The application software for a WSN is still undergoing development as well. In
fact, there is usually not a single part to a WSN that can be designed separately
without considering the whole system. This is often regarded as cross-layer
optimization.

When parts of an application are being executed in the sensor nodes, several
considerations must be taken into account. Of course, the storage, computation, and communication parts of the application should be balanced in terms
of power consumption, yet this is not the only issue to consider. For instance, an
application that is designed assuming completely reliable communications is
probably going to fail when deployed in a real WSN. The middleware layer of
the WSN may offer several QoS possibilities for the application, but total
reliability or extremely short and bounded delays are not realistic and will
probably not be available due to the inherent restrictions of the network. This
has to be taken into account when devising the algorithms and procedures for
the services and applications.

One example of application functionality that is part of a current open
research issue is the WSN's tracking of multiple targets. There is a trade-off
between the reliability, timeliness, and precision of the tracking information
and energy consumption. Several research papers deal with target tracking: See,
for instance, Lesser et al. (2003), in which specialized agents perform the targets
tracking, and Arora et al. (2004), who present a tracking system that is capable
of following several targets with high precision and employs a high number of
sensor nodes (78) in a very restricted area (approx. 18 × 8 meters).

4.5.6 The Most Important Innovations Considering
the Application Scenarios

Since this book was written mainly in the context of a European research
project, we made the effort of identifying which of the aforementioned major
open issues could be demonstrated in the application scenarios to be deployed
in Birstonas, Lithuania (see Chapter 8 for a detailed description). The fields in
which some of the most interesting innovations can be included, while taking
into account the characteristics of the scenarios, are summarized as follows:

- Middleware architecture for wireless sensor networks. In regard to the low-
  level detail abstraction to be provided for the applications, one of the results
  of the project will be a middleware architecture and implementation plan for
  the WSN. This middleware has to be general enough to be able to be used for

the different applications that will coexist in the Birstonas demonstrator. For instance, it has to support both event-based application functionalities with delay requirements such as in the surveillance scenario, and periodic and continuous monitoring of information with less strict delay requirements, such as the environmental data monitoring.

- Advanced issues in WSN middleware design. Some middleware issues that are no longer a problem in traditional wired networks become difficult to solve when dealing with WSNs. Among them the following can be mentioned: Offering localization transparency of the application functionality is not a trivial matter when the sensor nodes can lack an individual identifier. Likewise, if the component communication model is to provide a certain degree of reliability, the possible unreliability of the network has to be somehow compensated for. Additionally, a certain degree of functional self-configuration of the middleware could be valuable, for instance, to support different sensor configurations or even the remaining energy level of the node.

- Agent paradigm for WSNs. One of the intended results of the project is the design and implementation of an agent architecture that is part of the middleware. Thus, the open issues regarding the use of multi-agent approaches in WSNs are also of great importance for the uSWN project.

- Quality-of-service support. The middleware mechanisms for supporting the QoS that the applications require will also have to be considered in the project. Many of the application functionalities selected for the scenarios require reliability, bounded delays, or both.

- Multiple-target tracking functionality. One of the selected application functionalities for the Birstonas scenarios is precisely multiple-target tracking. As stated above, obtaining precise results with a reasonable number of sensor nodes is a technological challenge of potentially great complexity.

References

Abdelzaher T, Blum B, Cao Q, et al. (2004) EnviroTrack: Towards an environmental computing paradigm for distributed sensor networks. In *Proceedings of the 24th IEEE International Conference on Distributed Computing Systems (ICDCS'04)*, pp. 582–589.

Arora A, Dutta P, Bapat S, et al. (2004) A line in the sand: A wireless sensor network for target detection, classification, and tracking. *Comp Networks* 46(5):605–34.

atemu—Sensor network emulator/simulator/debugger. http://www.isr.umd.edu/CSHCN/research/atemu. Accessed December 2007.

UCLA Compilers Group (2005) Avrora: The AVR simulation and analysis network. http://compilers.cs.ucla.edu/avrora/. Accessed December 2007.

Bajaj L, Takai M, Ahuja R, et al. (1999) GloMoSim: A scalable network simulation environment. UCLA Computer Science Department Technical Report 990027.

Bakshi A, Prasanna VK, Reich J, et al. (2005) The Abstract Task Graph: A methodology for architecture-independent programming of networked sensor systems. In *Proceedings of the 2005 Workshop on End-to-End, Sense-and-Respond Systems, Applications and Services (EESR '05)*, pp. 19–24.

Balani R, Han CC, Rengaswamy RK, et al. (2006) Multi-level software reconfiguration for sensor networks. In *Proceedings of the 6th ACM & IEEE International Conference on Embedded Software (EMSOFT'06)*, pp. 112–121.

Baldwin P, Kohli S, Lee EA, et al. (2004) Modeling of sensor nets in Ptolemy II. In *Proceedings of the 3rd International Symposium on Information Processing in Sensor Networks (IPSN 2004)*, pp. 359–368.

Barr R, Bicket JC, Dantas DS, et al. (2002) On the need for system-level support for ad hoc and sensor networks. *Oper Syst Rev* 36(2):1–5.

Blumenthal J, Handy M, Golatowski F, et al. (2003) Wireless sensor networks—New challenges in software engineering. In *Proceedings of the 2003 IEEE Conference on Emerging Technologies and Factory Automation (ETFA '03)*, pp. 551–556.

Blumenthal J, Reichenbach F, Golatowski F, et al. (2005) Controlling wireless sensor networks using SeNeTs and EnviSense. In *Proceedings of the 3rd IEEE International Conference on Industrial Informatics (INDIN '05)*, pp. 262–267.

Buonadonna P, Gay D, Hellerstein JM, et al. (2005) TASK: Sensor network in a box. In *Proceedings of the 2 nd European Workshop on Wireless Sensor Networks*, pp. 133–144.

Cornell University. COUGAR: The network is the database. http://www.cs.cornell.edu/database/cougar/. Accessed October 2006.

Crossbow Technology Inc. Wireless systems for environmental monitoring. http://www.xbow.com/Products/Product_pdf_files/Wireless_pdf/Smart_Dust_AppNote.pdf. Accessed October 2006.

Crossbow Technology, Inc. (2007) MoteView User's Manual. http://www.xbow.com/Support/Support_pdf_files/MoteView_Users_Manual.pdf. Accessed December 2007.

Demmer M, Levis P (2004) Tython: Scripting TOSSIM. http://www.tinyos.net/tinyos-1.x/doc/tython/manual.html. Accessed December 2007.

eCos Home page. http://ecos.sourceware.org/. Accessed December 2007.

e-SENSE Project Deliverable D2.2.1 (2006) Initial e-SENSE system architecture.

EYES Project Deliverable 1.1 (2002) System architecture specification.

EYES Project Deliverable 3.3 (2003) Semantic addressing.

Fok CL, Roman GC, Lu C (2005) Rapid development and flexible deployment of adaptive wireless sensor network applications. In *Proceedings of the 25th IEEE International Conference on Distributed Computing Systems (ICDCS'05)*,. pp. 653–662.

Gay D, Levis P, von Behren R, et al. (2003) The nesC language: A holistic approach to networked embedded systems. In *Proceedings of ACM SIGPLAN 2003 Conference on Programming Language Design and Implementation*, pp. 1–11.

Girod L, Ramanathan N, Elson J, et al. (2007) Emstar: A software environment for developing and deploying heterogeneous sensor-actuator networks. *ACM Trans Sensor Networks* 3(3).

UCLA Parallel Computing Laboratory (2001) GloMoSim Global Mobile Information Systems Simulation Library. http://pcl.cs.ucla.edu/projects/glomosim/. Accessed December 2007.

Gold R (2005) Janus: An architecture for flexible access to sensor networks. In *Proceedings of the 2005 ACM Conference on Emerging Network Experiment and Technology*, pp. 248–249.

Gu L, Stankovic JA (2006) t-kernel: Providing reliable OS support to wireless sensor networks. In *Proceedings of the 4th International Conference on Embedded Networked Sensor Systems*, pp. 1–14.

Gummadi R, Gnawali O, Govindan R (2005) Macro-programming wireless sensor networks using Kairos. In *Proceedings of the 2005 International Conference on Distributed Computing in Sensor Systems (DCOSS 05)*.

Hadim S, Mohamed N (2006) Middleware: Middleware challenges and approaches for wireless sensor networks. *IEEE Distrib Syst Online* 7(3).

Han Q, Venkatasubramanian N (2001) AutoSeC: An integrated middleware framework for dynamic service brokering. *IEEE Distrib Syst Online* 2(7).

Han CC, Kumar R, Shea R, et al. (2005) A dynamic operating system for sensor nodes. In *Proceedings of the 3rd International Conference on Mobile Systems, Applications, and Services*, pp. 163–176.

Hayes-Roth B (1995) An architecture for adaptive intelligent systems. *Artif Intell* 72(1–2): 329–65.

J-Sim Home page. http://www.j-sim.org/. Accessed December 2007.

Lesser V, Ortiz CL Jr, Tambe M (Eds) (2003) *Distributed Sensor Networks: A Multiagent Perspective*. Kluwer Academic, New York.

Levis P, Culler D (2002) Maté: A tiny virtual machine for sensor networks. In *Proceedings of the 10th International Conference on Architectural Support for Programming Languages and Operating Systems*, pp. 85–95.

Levis P, Lee N (2003) TOSSIM: A simulator for TinyOS networks. http://www.cs.berkeley. edu/~pal/pubs/nido.pdf. Accessed October 2006.

Levis P, Lee N, Welsh M, et al. (2003) TOSSIM: Accurate and scalable simulation of entire TinyOS applications. In *Proceedings of the 1st International Conference on Embedded Networked Sensor Systems*, pp. 126–137.

Li S, Lin Y, Son SH, et al. (2003) Event detection services using data service middleware in distributed sensor networks. In *Proceedings of the 2nd International Workshop on Information Processing in Sensor Networks (IPSN '03)*.

Liu T, Martonosi M (2003) Impala: A middleware system for managing autonomic, parallel sensor systems. In *Proceedings of the 9th ACM SIGPLAN Symposium on Principles and Practice of Parallel Programming*, pp. 107–118.

Liu J, Perrone LF, Nicol DM, et al. (2001) Simulation modeling of large-scale ad-hoc sensor networks. In *Proceedings of the 2001 European Simulation Interoperability Workshop*.

Madden S, Franklin MJ, Hellerstein JM, et al. (2002) TAG: A Tiny AGgregation service for ad-hoc sensor networks. Oper Syst Rev 36 *(Special Issue: Physical Interface)*:131–46.

Madden S, Hellerstein J, Hong W (2003) TinyDB: In-network query processing in TinyOS, Version 0.4. http://telegraph.cs.berkeley.edu/tinydb/tinydb.pdf. Accessed December 2007.

MANTIS Project Home page. http://mantis.cs.colorado.edu/. Accessed December 2007.

Murphy AL, Heinzelman WB (2002) MiLAN: Middleware linking applications and networks. Technical Report: TR795. University of Rochester, Rochester, NY.

Newton R, Welsh M (2004) Region streams: Functional macroprogramming for sensor networks. In *Proceedings of the 1st International Workshop on Data Management for Sensor Networks*, pp. 78–87.

ns-2 main page. http://nsnam.isi.edu/nsnam/index.php/Main_Page. Accessed December 2007.

OMNeT++ discrete-event simulation system. http://www.omnetpp.org/. Accessed December 2007.

Technologies, Inc. http://www.opnet.com. Accessed December 2007.

Park S, Savvides A, Srivastava MB (2001) Simulating networks of wireless sensors. Winter Simulation Conference 2001, Vol. 2, pp. 1330–1338.

UCLA Parallel Computing Laboratory (2001) PARSEC: Parallel simulation environment for complex systems. http://pcl.cs.ucla.edu/projects/parsec/. Accessed December 2007.

Polley J, Blazakis D, McGee J, et al. (2004) ATEMU: A fine-grained sensor network simulator. In *Proceedings of the 1st Annual IEEE Communications Society Conference on Sensor and Ad Hoc Communications and Networks (IEEE SECON 2004)*, pp. 145–152.

Department of EECS, UC Berkeley (2007) The Ptolemy project. http://ptolemy.eecs.berkeley. edu. Accessed December 2007.

Scalable Network Technologies, Inc. (2007) QualNet 4.0 documentation. http://www. scalable-networks.com/publications/documentation/index.php. Accessed December 2007.

Ribeiro ARL, Silva FCS, Freitas LC, et al. (2005) SensorBus: A middleware model for wireless sensor networks. In *Proceedings of the 3rd International IFIP/ACM Latin American Conference on Networking*, pp. 1–9.

Römer K, Frank C, Marrón PJ, et al. (2004) Generic role assignment for wireless sensor networks. In *Proceedings of the 11th ACM SIGOPS European Workshop.*

Russell SJ, Norvig P (1995) *Artificial Intelligence: Modern Approach.* Prentice Hall, Englewood Cliffs, NJ.

Schriber TJ, Brunner DT (2006) Inside discrete-event simulation software: How it works and why it matters. In *Proceedings of the Winter Simulation Conference 2006 (WSC 06)*, pp. 118–128.

SENS: A sensor, environment and network simulator. http://osl.cs.uiuc.edu/sens/. Accessed December 2007.

SensorSim: A simulation framework for sensor networks. http://nesl.ee.ucla.edu/projects/sensorsim/. Accessed December 2007.

Sharifi M, Taleghan MA, Taherkordi A (2006) A middleware layer mechanism for QoS support in wireless sensor networks. In *Proceedings of the 2006 International Conference on Networking, International Conference on Systems and International Conference on Mobile Communications and Learning Technologies 2006 (ICN/ICONS/MCL 2006).*

Shen CC, Srisathapornphat C, Jaikaeo C (2001) Sensor information networking architecture and applications. *IEEE Pers Commun* 8(4):52–9.

Simon D, Cifuentes C, Cleal D, et al. (2006) Java™ on the bare metal of wireless sensor devices: The Squawk Java virtual machine. In *Proceedings of the 2nd International Conference on Virtual Execution Environments,* pp. 78–88.

Disco Lab, Rutgers University. Smart Messages. http://discolab.rutgers.edu/sm/. Accessed October 2006.

Computer Science Department, Cornell University. SNS: A staged network simulator. http://www.cs.cornell.edu/People/egs/sns/. Accessed December 2007.

Sobeih A, Hou JC, Kung LC, et al. (2006) J-Sim: A simulation and emulation environment for wireless sensor networks. *IEEE Wirel Commun* 13(4):104–9.

UCLA NESL. SOS 2.x Home page. https://projects.nesl.ucla.edu/public/sos-2x/doc/. Accessed December 2007.

Souto E, Guimarães G, Vasconcelos G, et al. (2004) A message-oriented middleware for sensor networks. In *Proceedings of the 2nd Workshop on Middleware for Pervasive and Adhoc Computing,* pp. 127–134.

Souto E, Guimarães G, Vasconcelos G, et al. (2005) Mires: A publish/subscribe middleware for sensor networks. *Pers Ubiquit Comput* 10(1):37–44.

St Ville L, Dickman P (2003) Garnet: A middleware architecture for distributing data streams originating in wireless sensor networks. In *Proceedings of the 23rd International Conference on Distributed Computing Systems Workshops,* pp. 235–240.

Crossbow Technology Inc. Stargate NetBridge. http://www.xbow.com/Products/productdetails.aspx?sid=275. Accessed December 2007.

Sundresh S, Kim W, Agha G (2004) SENS: A sensor, environment and network simulator. In *Proceedings of the 37th Annual Symposium on Simulation.*

Sun Microsystems, Inc. Project SUN Spot. http://www.sunspotworld.com/. Accessed November 2006.

JiST/SWANS: Java in Simulation Time/Scalable Wireless Ad hoc Network Simulator. http://jist.ece.cornell.edu/. Accessed December 2007.

Szumel L, LeBrun J, Owens JD (2005) Towards a mobile agent framework for sensor networks. In *Proceedings of the 2nd IEEE Workshop on Embedded Networked Sensors (EmNetS-II),* pp. 79–88.

Titzer BL, Lee DK, Palsberg J (2005) Avrora: Scalable sensor network simulation with precise timing. In *Proceedings of the 4th International Symposium on Information Processing in Sensor Networks (IPSN 2005),* pp. 477–482.

Turon M (2005) MOTE-VIEW: A sensor network monitoring and management tool. In *Proceedings of the 2nd IEEE Workshop on Embedded Networked Sensors (EmNetS-II),* pp. 11–18.

Varga A (2001) The OMNeT++ discrete event simulation system. In *Proceedings of the European Simulation Multiconference (ESM'2001)*.

Walsh K, Sirer EG (2004) Staged simulation: A general technique for improving simulation scale and performance. *ACM Trans Model Comput Simul* 14(2):170–95.

Welsh M, Mainland G (2004) Programming sensor networks using abstract regions. In *Proceedings of the 1st Conference on Symposium on Networked Systems Design and Implementation,* Vol. 1.

Whitehouse K, Zhao F, Liu J (2005) Semantic streams: A framework for declarative queries and automatic data interpretation. Microsoft Research Technical Report MSR-TR-2005-45.

Wolenetz M, Kumar R, Shin J (2005) A simulation-based study of wireless sensor network middleware. *Int J Netw Manag* (Online) 15(4):255–67.

Yannakopoulos J, Bilas A (2005) CORMOS: A communication-oriented runtime system for sensor networks. In *Proceedings of the 2nd European Workshop on Wireless Sensor Networks,* pp. 342–353.

Ye W, Heidemann J, Estrin D (2002) An energy-efficient MAC protocol for wireless sensor networks. In *Proceedings of the 21st Annual Joint Conference of the IEEE Computer and Communications Societies (INFOCOM 2002),* Vol. 3,. pp. 1567–1576.

Yu Y, Krishnamachari B, Prasanna VK (2004) Issues in designing middleware for wireless sensor networks. *IEEE Netw* 18(1):15–21.

Chapter 5
Network Aspects and Deployment in WSNs

Abstract Wireless sensor networks have been found to be very useful for many military and civil applications such as disaster management, surveillance of battle fields, and security. In many of these environments, the sensor nodes are strongly limited in terms of energy since their batteries usually cannot be recharged. Thus, designing energy-efficient algorithms has become an important factor to lengthen the lifetime of WSNs. Efficient network deployment and management is crucial to set an acceptable quality level in the network operation and to preserve as much of the node energy as possible. Correct energy management assures the desired performance level for data transmissions while lengthening the lifetime of the network. Energy restrictions combined with wide-scale deployments make implementing energy-saving methods necessary in most protocols, including the network and MAC layers. Energy-efficient routing can optimize the lifetime of the network by selecting paths that expend less energy, whereas collision suppression and decreasing energy consumption in the receiver must be the goals of the different Medium Access Control (MAC) mechanisms. Since energy considerations have dominated most of the investigations about WSN network operation and deployment, quality-of-service (QoS) issues such as latency, throughput, delay, or jitter have not been treated with great detail until now, topics that have been identified as interesting open issues for further research.

5.1 WSN Topologies and Deployment Methodologies

The term "topology" refers to the physical disposition in which the nodes of a network (in this case, a WSN) are connected to one another. Network topology only refers to node connections. The distance between nodes, physical interconnections, transmission rates, or types of signals do not belong in this category, although they can be influenced by the topology. However, a good WSN design takes the topology into account when improving several performance factors such as energy efficiency, robustness, or general QoS parameters.

To understand the topologies of a WSN, the types of nodes that form the network first need to be introduced. A WSN contains both sources and sinks.

A.-B. García-Hernando et al., *Problem Solving for Wireless Sensor Networks*,
DOI: 10.1007/978-1-84800-203-6_5, © Springer-Verlag London Limited 2008

A source can be any entity in the network that is able to provide information. It is usually a sensor node, but it can also be an actuator node that provides feedback about an operation (Karl and Willig, 2005). On the other hand, a sink is the entity that requires information. There are two possibilities for a sink: It can belong to the WSN and be just another sensor/actuator node, or it can be an external entity. If the sink is an actuator belonging to the WSN, it could be, for example, a laptop used to interact with the sensor nodes. If it is an external element, the sink may be a gateway to another network such as the Internet, where the information requests come from some external device/node indirectly connected to the WSN. These main types of sinks are illustrated in Fig. 5.1.

The types of network topologies can be classified according to several criteria. In addition, the network hierarchy should be taken into account when selecting a suitable and efficient routing scheme. In fact, the main WSN topology division is based on the existence or absence of hierarchy among network elements.

In flat networks—or those networks without hierarchy—each node has the same capabilities. Thus, control over the routes and channels must be performed in a distributed fashion.

In hierarchical networks, some nodes will have different capabilities than others. These capabilities are divided into two areas: physical, where the nodes or links have different physical characteristics, and logical, in which the nodes have different functions in the network.

The most common and representative WSN topologies are the following (Fig. 5.2):

- Ad hoc without hierarchy: In this case, all nodes are equal. They are their own service providers, and thus data pass from node to node to reach a sink.

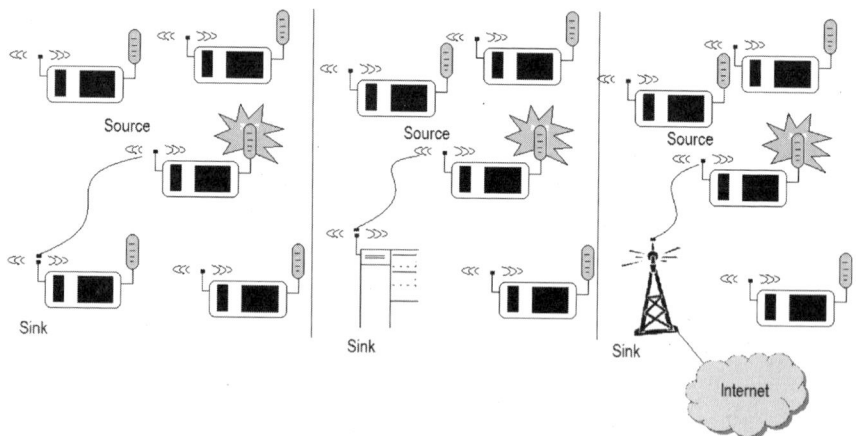

Fig. 5.1 Three cases to illustrate the types of sinks. (From [Karl and Willig, 2005]. ©John Wiley & Sons Ltd. Reproduced with permission.)

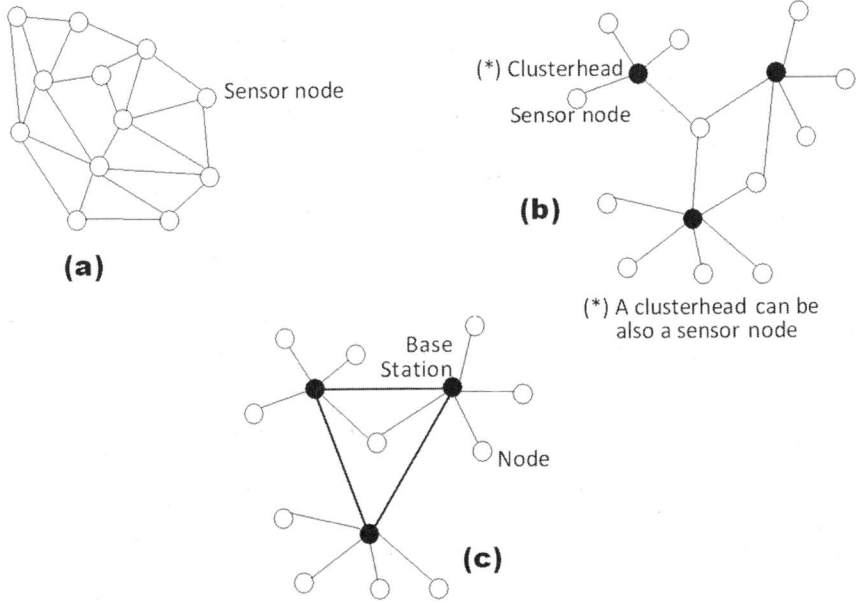

Fig. 5.2 Some network topologies: (a) ad hoc, (b) clustered, and (c) overlay. (From [Pottie and Kaiser, 2005].)

A common example of this type of network is the mobile ad hoc networks (MANETs), although this scheme can also be valid for networks formed by nodes of low or no mobility.

- Hierarchical network by clustering: The idea of hierarchy implies assigning some nodes with a special role, for example, controlling neighboring nodes. In this sense, local groups or clusters can be formed; the "controllers" of such groups are often referred to as cluster heads. The major functions of the cluster heads are local resource arbitration (i.e., in MAC protocols), making routing tables more stable since all traffic is routed through the cluster heads, and making higher-layer protocols more scalable since the higher layer perceives a less complex network due to clustering. Furthermore, cluster heads are the usual places where the traffic is aggregated and compressed to converge to a single sink.

- Overlay networks: This type of clustering network has both physical and logical hierarchies. Nodes that assume special control functions are thus more powerful and/or have privileged capacities with respect to the rest. This way the more powerful nodes may form a network on their own, allowing higher scalability. An example of this type of network is a cellular network, where cells are controlled by base stations, which, in turn, are connected to a wired infrastructure for intercell routing.

Another possible topology in a network is called a mesh topology. In this case, all sensor nodes are identical and can communicate directly with each

Fig. 5.3 Multi-hop
networks: When direct
communication is
impossible due to the
distance and/or obstacles,
multi-hop communication
can be a solution. (From
(Karl and Willig, 2005).
©John Wiley & Sons Ltd.
Reproduced with
permission.)

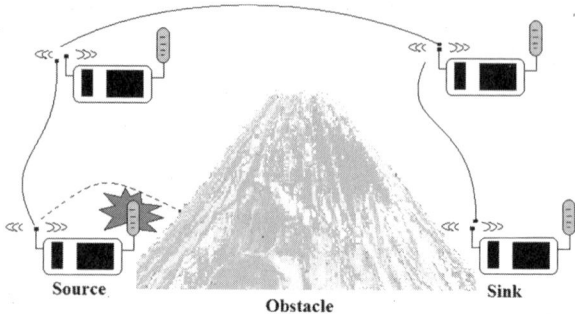

other, providing a high level of redundancy in the data paths between nodes. In a mesh WSN, every node should be in the area of radio coverage of any other node, which is a disadvantage since nodes in a WSN have limited power. This reduced available energy makes it unviable to implement a WSN with mesh topology if the coverage area exceeds certain dimensions, for example in environmental or agriculture applications, or if it has a strong attenuation, such as inside buildings. Therefore, in many situations it is necessary to accept a topology that is not completely meshed, having some nodes route the information of others without being the packet destination. This is known as multi-hop routing (see Fig. 5.3)

A topology with multi-hop routing has both advantages and disadvantages with respect to a meshed topology. Some advantages are the following:

- Not only is the multi-hop routing a functional solution for solving problems with large distances or obstacles, but it also has been used for improving energy efficiency in communications. The radio channel attenuation increases at least at a quadratic rate with the distance in most environments, thus wasting less energy with a multi-hop architecture than with single-hop topologies. The global power consumption is lower if the nodes transmit to other neighboring nodes than in a hypothetical situation in which every node transmits directly to a sink or gateway.
- If the density of intermediate nodes (relays of the information) is larger, the reutilization frequency distance is shorter. Therefore, the global capacity for data transmissions increases.
- For several applications it is very convenient to carry out data aggregation in the intermediate nodes instead of transmitting all the raw data generated by nodes. A multi-hop routing allows the nodes that route the information to aggregate the data received with their own data and transmit only the summarized or aggregated information. Sending less information increases both the global information transmission capacity of the network and its lifetime, thus saving energy.

Among the disadvantages of multi-hop routing are (1) the larger delay between generating the information and its reception by the sink and (2) some applications' requirement for controlled delays. The reasons for these issues are the following (Pottie and Kaiser, 2005):

- Each packet will be queued inside each of the nodes through which it is routed, producing larger and, in general, variable delays. If the percentage of resource use in the WSN is low, this delay may not be significant, although the application determines what it significant and what is not.
- Even if queuing delays are not significant, the functioning of the MAC protocol in each hop may add an important amount of time. For instance, in many MAC protocols, the nodes switch from states of low or null activity with very low energy consumption to activity states in which they may send or receive data. Waiting for an active period of a neighboring node in order to send messages may be a source of delay. Another example can be found in MAC protocols designed for star topologies, in which a node must join the master of the neighboring star nearest the destination prior to forwarding a packet. Allowing a node to be in the influential area of more than one master node may alleviate this problem.

In consideration of the physical deployment of sensor networks in the field, Akyildiz et al. (2002) proposed a strategy for WSN deployment and topology maintenance comprising the following three phases: (1) the pre-deployment and deployment phase where sensor nodes can be either thrown in mass or placed one by one in the sensor field; (2) the post-deployment phase where topology changes are taken into account both for stationary WSNs and for mobile sensor nodes. These changes are due to changes in sensor node position, a changing radio range, available energy, malfunctioning and task details; (3) the redeployment of additional nodes phase. This phase can be carried out at any time during the WSN life, to replace the malfunctioning nodes or due to changes in task dynamics.

5.1.1 Self-Organization

Some optimization factors in a WSN such as scalability or robustness require the network to be organized in a distributed fashion. This means that there should not be any responsible centralized entity, i.e., an entity controlling the medium access or making routing decisions, doing the same tasks as a base station in a cellular mobile network. The disadvantages of a centralized approach are the unique path (only one failure point) and the difficulty of its implementation in a radio network, where the participants have a limited communication range. Rather, the nodes in WSNs should organize the network cooperatively by using distributed algorithms and protocols. "Self-organization" is the term commonly used to name this approach.

However, organizing a network in a distributed fashion is not free of potential failures. In many circumstances, a centralized approach could produce solutions that perform better or require fewer resources, in particular, energy. A centralized mechanism using dynamic selection of specific nodes between a set of identical nodes, which assume the responsibilities of a centralized agent, can be adopted to exploit the advantages of using a centralized approach. This election generates a hierarchy and must be a dynamic selection. The election process should be repeated continuously before overloading the resources of the selected nodes and thus wasting their energy reserves. The disadvantage of low robustness is inherent to hierarchical networks. The election of rules and triggering conditions for reselection vary significantly depending on the applications for which these hierarchies are used.

The major tasks carried out in the network self-organization process are establishing topology (connectivity discovery), channel assignment in the case of wireless nodes to avoid collisions, and connectivity maintenance (Pottie and Kaiser, 2005).

Discovering neighboring nodes can be achieved by hand-shaking procedures, preferably using low-power transmissions to establish communication only with the neighboring nodes. Usually, WSN design is intended to save energy rather than to take advantage of the spectrum. Spatial reuse of channels is possible when considering that the range of a node transmitter is limited. Therefore, synchronism mechanisms can be established for a local group of nodes forming clusters, without needing global network synchronization. These mechanisms are used often by MAC protocols designed expressly for WSNs, some of which we describe below.

There are two main limitations in wireless links:

- Any transceiver needs a minimum signal strength to demodulate signals successfully.
- The received power decreases as the distance between the transmitting and receiving nodes increases.

These facts generate two common transmission problems in WSNs that should be solved during the connectivity discovery phase: hidden-terminal and exposed-terminal problems. The hidden-terminal problem occurs specifically with Carrier Sense Multiple Access (CSMA) protocols, where a node senses the medium before starting to transmit a packet. If the medium is busy, the node will discard the packet transmission to avoid a collision and a subsequent retransmission. Consider the example in Fig. 5.4. Three nodes, A, B, and C, are deployed in such a way that A and B are in mutual coverage. Nodes B and C are in mutual coverage as well, but A and C cannot hear each other. Thus, neither of the two nodes knows that the other exists. Assume that A starts to transmit a packet to B and later node C also decides to start a packet transmission. A carrier-sensing operation by C shows a free medium since C cannot hear signals from node A. When C starts to transmit the packet, the signals will produce a collision at node B and both packets will

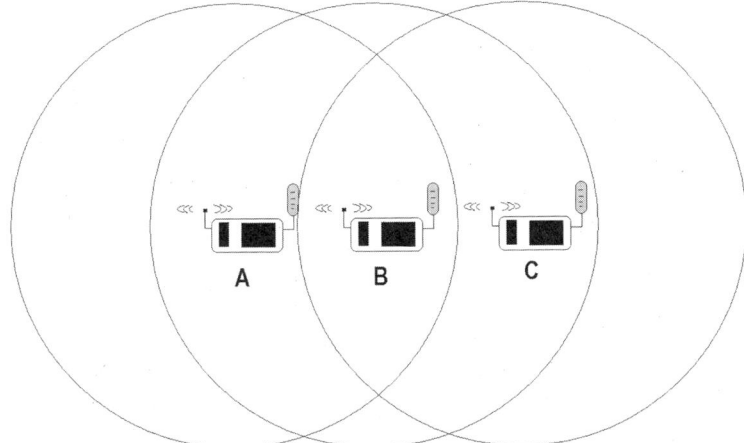

Fig. 5.4 Hidden-terminal scenario (circles indicate transmission and interference range)

be lost. Thus, with CSMA MAC protocols, the hidden-terminal scenario will surely produce collisions. The hidden-terminal problem can be avoided with the use, for instance, of the set of RTS/CTS primitives (Request to Send and Clear to Send).

In the exposed-terminal scenario (Fig. 5.5), node B transmits a packet to A, and a short time later, C wishes to transmit a packet to D. It can be expected that these actions take place simultaneously since theoretically there is no collision. Yet the carrier-sense operation performed by C suppresses C's transmission, wasting bandwidth.

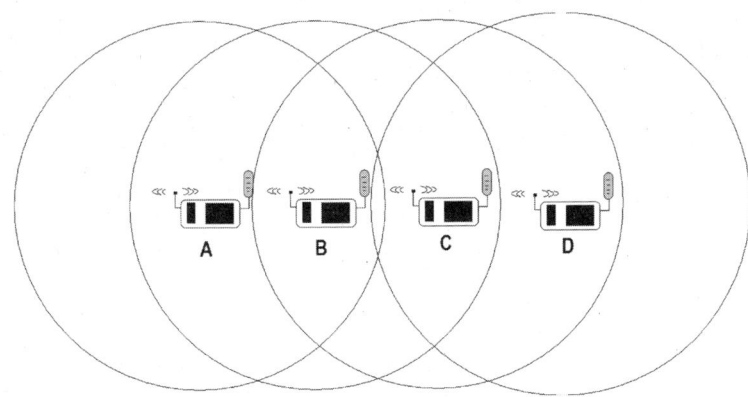

Fig. 5.5 Exposed-terminal scenario. (From [Karl and Willig, 2005]. ©John Wiley & Sons Ltd. Reproduced with permission.)

5.2 Communication Protocol Architectures

The protocol stack used in wireless sensor networks combines power and routing awareness, integrates data with networking protocols, communicates power efficiently through the wireless medium, and promotes cooperative efforts of sensor nodes. The protocol stack consists of the application layer, transport layer, network layer, data link layer, physical layer, power management plane, mobility management plane, and task management plane (see Fig. 5.6)

5.2.1 Physical Layer

The physical layer is the first level of the protocol stack. It performs services requested by the data link layer. The physical layer is the most basic network layer, providing only the means for transmitting raw bits rather than packets over a physical data link connecting network nodes. No packet headers or trailers are consequently added to the data by the physical layer. The bit stream may be grouped into code words or symbols and converted to a physical signal that is transmitted over a physical transmission medium, which is the wireless medium in a WSN. The physical layer provides an electrical, mechanical, and procedural interface to the transmission medium. Broadcast frequencies, the modulation scheme used, and similar low-level features are specified in the physical layer.

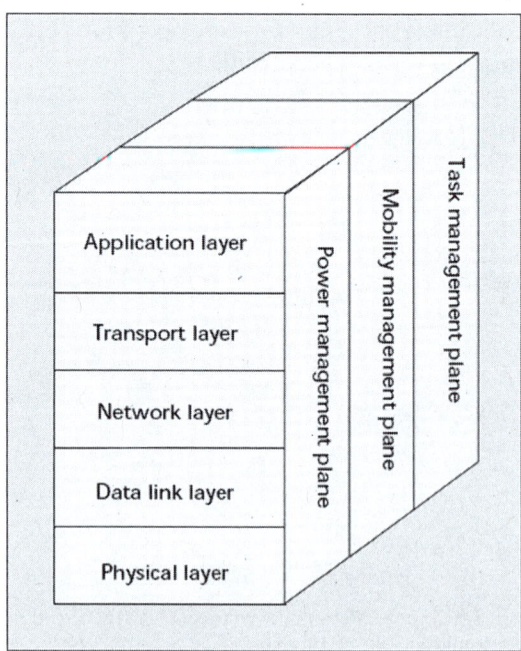

Fig. 5.6 WSN protocol stack. (From [Akyildiz et al., 2002]. ©2002 IEEE.)

The physical layer determines the bit rate, also known as the channel capacity, digital bandwidth, maximum throughput, or connection speed.

A variety of physical layer wireless transmission technologies are used in traditional wireless networks. Considering the specific physical-layer requirements of wireless sensor networks and taking into consideration the particular characteristics and usage scenarios, it can be inferred that spread-spectrum technologies meet the requirements much better than narrowband technologies. Furthermore, ultra wideband technologies are found to be a promising emerging alternative.

Wong (2004) considers the design of the physical layer in wireless sensor networks. Wong shows that compared to narrowband technologies, applying a spread spectrum in a physical layer has many advantages such as low power consumption, robustness to interference, ease of synchronization, and physical-layer multi-case ability. For those reasons, energy-efficient schemes in spread-spectrum systems should be investigated for WSNs.

5.2.2 Data Link Layer

The data link layer is responsible for multiplexing data streams, data frame detection, medium access, and error control. It ensures reliable point-to-point and point-to-multipoint connections in a communication network. The most important tasks of the link layer are the formation and maintenance of direct communication associations ("links") between neighboring nodes and the reliable and efficient transfer of information across these links. Reliability has to be achieved despite time-variable error conditions on the wireless link. Nevertheless, the collaborative and application-oriented nature of the sensor networks and the physical constraints of the nodes, such as energy and processing limitations, determine the way in which these responsibilities are fulfilled.

This layer is subdivided into Logical Link Control (LLC) and Medium Access Control (MAC). In WSNs the fundamental design issue is the MAC. MAC protocols solve a seemingly simple task of coordinating when a number of nodes access a shared communication medium. We will explain the specific requirements and problems of a WSN MAC layer and present the fundamental MAC protocols.

5.2.2.1 MAC Requirements for WSNs

Medium Access Control design in sensor networks is very different from traditional wireless MAC schemes due to the inherent WSN limitation, among them the energy constraint. The MAC protocol in a wireless multi-hop, self-organizing sensor network must achieve two main goals:

- Creating network infrastructure: Since thousands of sensor nodes are densely scattered in a sensor field, the MAC scheme should establish communication links for data transfer. This forms the basic infrastructure needed for wireless communication and gives the sensor network self-organizing ability.
- Efficiently using and sharing energy and communication resources between sensor nodes: Novel protocols and algorithms are needed to effectively tackle the unique resource constraints and application requirements of sensor networks, which means that MAC schemes in other wireless networks cannot be adopted into the sensor network scenarios. Mobility also poses unique challenges to MAC protocol design since weak mobility implies topology changes, while strong mobility means new nodes or node failures.

WSN requirements are different from those of traditional wireless networks. The additional requirements come principally from the need to save energy. The importance of energy efficiency for MAC protocols design is relatively new; thus, many of the classical protocols like ALOHA and CSMA (Carrier Sense Multiple Access) do not take this requirement into account. Some papers dealing with energy concerns in MAC protocols are Chen et al. (1998), Goldsmith and Wicker (2002), and Woesner et al. (1998). Other typical performance characteristics such as fairness, throughput, or delay have played a minor role in WSNs, yet recently they have been receiving more attention. Fairness is not an important factor since the nodes in a WSN usually do not compete for bandwidth, instead collaborating to achieve a common goal depending on the application. The delay optimization of access/transmission is treated as an issue that works against energy conservation, and throughput is not a very important issue for most of the applications.

In a WSN, scalability and robustness requirements are confronted with the frequent changes in the topology, which are generally produced by temporary power decreases in nodes, node mobility, new node deployment, or "death" of existing nodes. The need for scalability is evident when considering a very dense WSN with dozens or thousands of nodes.

Good collision management is also important since it can be useful for saving energy, both in transmission from the source node and in reception at the destination node. Collisions should be avoided by design (fixed assignments/TDMA or assignments under demand protocols) or by suitable collision suppression procedures to offset the hidden-terminal problem in CSMA protocols.

Low complexity must be fulfilled by the MAC protocol for WSNs related to energy savings. Because the nodes used in WSNs are simple, they should not consume an exceptional amount of resources such as memory, energy, or processing power. Accordingly, computationally expensive operations, such as complex scheduling algorithms, should be discarded.

Most of the MAC protocols are classified in two groups: contention-based or schedule-based. The difference is the number of contestants that have the option of transmitting to a node at a given instant:

- In contention-based protocols, any node can try to transmit with the risk of collisions. As all nodes have to contend for the communication channel, collisions are possible and are one of the major causes of energy inefficiency. Consequently, these protocols have several mechanisms to suppress collisions or to reduce the probability of occurrence. In a contention-based wireless sensor network, since nodes can directly transmit information to the base station at any time, idle listening can also occur. This is one of the main sources of energy waste in these networks since the nodes normally remain inactive for a long time without transmitting. The benefit of these protocols is their simplicity and robustness.
- In schedule-based or polling-based protocols, only one neighbor has the opportunity to transmit at any given time, thus eliminating collisions. These protocols usually have a TDMA component, which also provides an implicit mechanism of passive listening suppression. When a node knows the slots it has been assigned, it is sure that the communication, both transmission and reception, will only be produced at these slots; otherwise, the receptor can be deactivated. This scheme is much more complicated since the base station must poll the nodes and then give each one a time to transmit. The constraint of these protocols is the large amount of data transmitted to set up the network structure. However, once the structure is created, there is no chance of collisions and nodes can save energy in their operation.
- In hybrid protocols, a combination of the previous protocol types is used.

5.2.2.2 Medium Access Protocols for WSNs

Several authors have suggested medium access schemes for WSNs, some of which are modifications of existing protocols for wireless ad hoc networks. This is still a growing area of research calling attention to several open issues yet to be addressed. Several recently proposed schemes are discussed below.

Contention-Based Protocols

Sensor MAC (S-MAC) (Ye et al., 2002; 2004)

S-MAC is a contention-based MAC protocol explicitly designed for wireless sensor networks. While reducing energy consumption is the primary goal in those networks, this protocol has also achieved good scalability and collision avoidance by using a combined scheduling and contention scheme.

To achieve the primary goal of energy efficiency, the main sources that cause the inefficient use of energy as well as what trade-offs can be made to reduce energy consumption need to be identified. In this way, the following major sources of energy waste are identified:

- Collisions: When a transmitted packet is corrupted, it has to be discarded; the follow-up retransmissions increase energy consumption. Collisions not only waste energy, but they increase latency as well.

- Overhearing: A node can pick up packets intended for other nodes.
- Control packet overhead: Sending and receiving control packets also consumes energy.
- Idle listening: Listening to receive possible traffic that was not sent can be the biggest cause of inefficiency, especially in many sensor network applications when nodes are in the idle state most of the time. Most sensor networks are designed to operate over a long period of time; since the nodes are idle for a long time, idle listening can be a dominant factor behind energy waste in such cases.

S-MAC tries to reduce energy waste from the above sources; in exchange, it accepts some reduction in both per-hop fairness and latency.

S-MAC establishes a low-duty-cycle operation of nodes in a multi-hop ad hoc network and reduces idle listening by periodically putting nodes into a sleep state. In the sleep mode, the radio is completely turned off. To reduce the control overhead and latency, S-MAC introduces coordinated sleeping among neighboring nodes. The use of in-channel signaling to put each node to sleep when its neighbor is transmitting to another node avoids the overhearing problem without requiring an additional channel. In S-MAC, the low-duty-cycle mode is the default operation of all nodes in the network. Nodes only become more active by changing the duty cycle when heavy traffic is present in the network or once an event occurs in the case of an event-driven WSN. S-MAC adopts a periodic wakeup scheme in which each node alternates between a fixed-length listen period and a fixed-length sleep period according to its schedule (Fig. 5.7). RTS and CTS primitives (Request to Send and Clear to Send) are used to avoid collisions (the hidden-terminal problem).

An important feature of WSNs that must be highlighted is the in-network data processing, which can greatly reduce energy consumption compared to transmitting all raw data to the end node. Therefore, in-network processing requires store and forward processing of messages, where a message is considered to be a meaningful unit of data that a node can process, average, filter, and transmit. Messages may be long and may consist of many small fragments. In

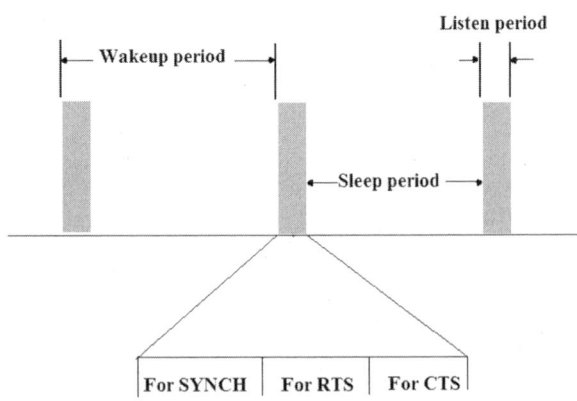

Fig. 5.7 S-MAC principle

such a case, MAC protocols that promote fragment-level fairness actually increase message-level latency for the application, whereas message passing reduces message-level latency by trading off the fragment-level fairness. In traditional wireless voice or data networks, each user desires equal opportunity and time to access the medium, so per-hop MAC-level fairness is thus an important issue. However, in sensor networks, all nodes work together toward a single common task and, at any particular time, one node may have more data to send than other nodes. In this case, fairness is not important as long as application-level performance is not degraded. In the S-MAC protocol, the concept of message passing helps efficiently transmit very long messages by dividing these messages into small fragments and transmitting them in a burst. The result is that a node with more data to send gets more time to access the medium. Message passing can be one way of saving energy by reducing control overhead and avoiding overhearing. It is also well suited to applications where nodes support in-network data processing since motes usually need to receive the complete message before they can begin to process the data.

X-MAC (Buettner et al., 2006)

X-MAC uses a duty-cycle mechanism similar to S-MAC in which the sender also begins with a preamble that the receiver will eventually hear when it wakes. However, X-MAC introduces some improvements over other duty-cycle MAC schemes. First, the sender does not emit a long preamble, but rather a series of shorter preambles (the authors call this a "strobed preamble") in which there is information about the receiver of the information that is going to be transmitted. Between the short preambles is enough time for the receiver to send an early acknowledgment as soon as the intended receiver is awake, which will make the sender stop emitting the preambles and start transmitting the information. This shortens the average time during which the sender emits the preamble and prevents other non-receivers from staying awake until the preamble ends.

Second, the authors also describe a means for automatically adapting the duty-cycle parameters to accommodate them to the traffic load conditions.

The benefits of this MAC scheme over other duty-cycle approaches are lower energy consumption (shorter preambles, shorter useless listening or overhearing), lower delays (shorter preambles before transmitting the information), and better scalability. Furthermore, this MAC can be implemented on top of all types of digital radios, whether bit stream–based or the more recent packet-based ones. The authors have implemented X-MAC on top of the Mantis Operating System (for which there are ported versions working in the MICA2, MICAZ, and TelosB platforms) and have tested their approach using TelosB sensor nodes.

Schedule-Based Protocols

Self-Organizing MAC for Sensor Networks (SMACS)

The Self-Organizing Medium Access Control for Sensor Networks (SMACS) protocol described by Sohrabi and Pottie (1999) and (Sohrabi et al., 2000) is part of a wireless sensor network protocol suite that addresses MAC, neighbor discovery, attachment of mobile nodes, a multi-hop routing protocol, and a local routing protocol for cooperative signal processing purpose. SMACS was designed for network startup and link-layer organization in a static WSN. SMACS essentially combines neighborhood discovery and assignment of TDMA (Time Division Multiple Access) schedules to nodes. In this scheme, each node maintains a TDMA frame in which the node schedules different time slots to communicate with its known neighbors. During each time slot, that node talks only to one neighbor. To avoid interference between adjacent links, the protocol uses different frequency channels (FDMA, Frequency Division Multiple Access) or spread-spectrum codes (CDMA, Code Division Multiple Access). Although the frame structure is similar to a typical TDMA frame, it does not prevent two interfering nodes from accessing the medium at the same time. The actual multiple access is accomplished by FDMA or CDMA.

Traffic-Adaptive Medium Access protocol (TRAMA)

The Traffic-Adaptive Medium Access (TRAMA) protocol as presented by Rajendran et al. (2003) creates schedules allowing nodes to access a single channel in a collision-free manner. It uses traffic-based information to decide on schedules for individual nodes and thus is adaptive to network traffic. The protocol assumes that all nodes are time-synchronized and divides time into random access periods and scheduled-access periods. A random access period followed by a scheduled-access period is called a cycle. The duty cycle of switching between these states could be adjusted according to the application requirements and also according to the different network types. For stationary networks, the random access periods occur less frequently, and vice versa for highly dynamic networks.

Communication in TRAMA consists of three major components: the neighbor protocol (NP), the adaptive election algorithm (AEA), and the schedule exchange protocol (SEP). The NP is used to exchange one-hop neighbor information among neighbors and to gather two-hop topology information for each node in the network. Nodes always start in the random access mode with the neighbor protocol. During scheduled access, the AEA selects transmitters and receivers achieving collision-free transmission. The SEP is used to exchange traffic schedules among neighbor nodes during the scheduled access mode. These schedules contain the set of receivers for the traffic currently originating at the node and its scheduled transmission slots.

Clustering-Based Protocols

The cluster-based hierarchical structure is used to support scalability. As the number of sensors is increased, more clusters can be formed without increasing the processing or communication loads on individual cluster heads. The three levels in the hierarchical design of this architecture consist of a base station (a data sink) at the top level, cluster heads at the middle level, and the other sensors at the leaf level (Fig. 5.8).

Fig. 5.8 Clustering
organization

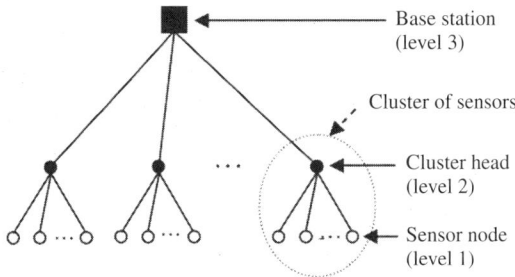

The base station is a machine capable of analyzing data collected from the cluster heads and displaying a global view of events being monitored. It is responsible for initiating and managing the network and is ultimately the gateway of the sensor network to the Internet or another destination. Sensors are deployed in large numbers across an area of observation. Their primary function is to collect data from their surroundings. Direct communication among the level 1 sensors occurs only at the time of cluster formation or cluster reconfiguration; otherwise, the main stream of communication consists of conveying data results to the corresponding cluster head. Before deployment, each sensor is given an ID (identifier) that uniquely identifies it and is used to authenticate data sent by the node. Cluster heads are selected from among the deployed sensors by a self-configuring mechanism that depends on the specific cluster-based protocol. Sensors in a particular cluster register themselves with their respective cluster head, which becomes the immediate point of contact for their sensors for communication and reporting purposes. The heads collect data from the sensors, aggregate the data, and send the results to the base station.

The deployment area is usually remote, the large number of sensor nodes to deploy for an application rules out manual configuration, and the environmental dynamic precludes design time pre-configuration. Therefore, nodes have to self-configure to establish a topology that enables communication and sensing coverage under stringent energy constraints. Clustering allows sensors to efficiently coordinate their local interactions in order to achieve global goals. Localization saves transmission energy since it allows communicating with a closer local coordinator instead of a more distant base station. It is well known that to transmit a signal over a distance d, the required radiation energy E is proportional to d to the power m, where m is 2 in the free space and ranges up to 4 in environments with multiple paths.

Another advantage of using clusters is the ability to perform data aggregation at the cluster heads, in which data collected from sensors are aggregated before being forwarded to the base station, which reduces the amount of data to transmit and thus saves power. Data aggregation is a paradigm for wireless sensor networks. The idea is to combine data from different sources and routes, eliminating redundancy, minimizing the number of transmissions, and saving energy. We must point out that sensor data are different from data associated with traditional wireless networks since it is not the data itself that is important, but rather the data analysis, which allows an end user to draw conclusions about the monitored environment. For example, if sensors are monitoring temperature, the end users could only be interested in a high-level description of the events occurring, such as minimum, maximum, or average temperatures. The main purpose of data aggregation is to reduce required communication and, in turn, the total energy consumption.

One example of a clustering-based or schedule-based MAC protocol for wireless sensor networks is LEACH.

LEACH (Heinzelman et al., 2000)

Low-Energy Adaptive Clustering Hierarchy, or LEACH, is a dynamic clustering method in which time is partitioned into intervals of equal length. At the beginning of the interval, each sensor becomes a cluster head with some predefined probability. That cluster head broadcasts messages to their neighbors; other sensors receive these messages and then join a cluster by choosing the cluster head with the strongest signal. During the communication interval cluster, members send information following a TDMA agenda (schedule or polling-based) to their cluster heads, which aggregate, compress, and route the information to the remote access point. Once the interval ends, the whole clustering process is restarted. The clusters and cluster heads are not fixed, and cluster heads consume more energy than cluster members in radio transmission. Therefore, rotating cluster heads is a way to distribute energy consumption evenly across all sensors in the network, which makes the sensor network last longer.

Wavenis MAC

The Wavenis MAC protocol, which may be considered a contention-, schedule-, and mesh cluster–based protocol, is responsible for

- Defining MAC addresses. The Wavenis device MAC addresses are coded on 6 bytes (48 bits) and stored in non-volatile memory at the factory.
- Coding and decoding Tx and Rx data frames.
- Managing Forward Error Correction with FEC coding that adds 1/3 data redundancy.
- Managing data scrambling by means of an LFSR feed with a pseudo-random sequence.
- Managing data interleaving through a 16×16 matrix.

Wavenis implements a sophisticated combination of fast FHSS (managed by a call to the PHY protocol), FEC, data scrambling, and data interleaving. The overall effect is equivalent to digital noise-spreading techniques where the bit stream is divided into 21-bit subpackets to which a BCH (31,21) error correction code is added, resulting in 32-bit subpackets. Eight successive subpackets are then scrambled, by means of an LFSR feed with a pseudo-random sequence, and interleaved through a 16×16 matrix. The resulting 256-bit stream delivered on

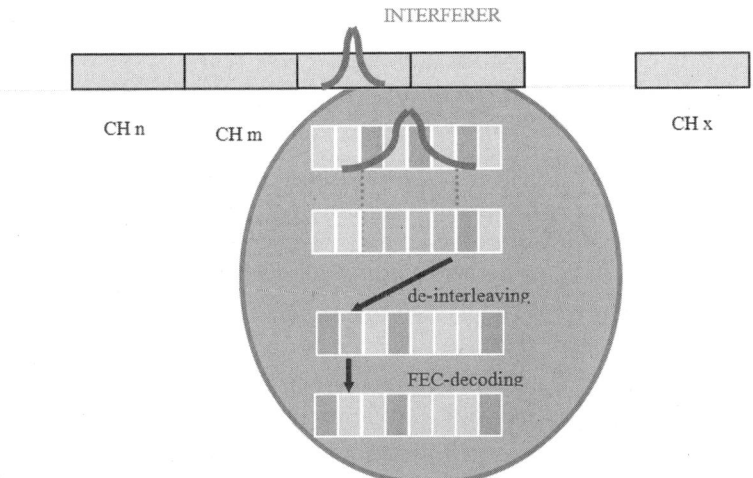

Fig. 5.9 The Wavenis MAC Forward Error Code (FEC) mechanism

matrix output contains effective data bits and FEC bits in a completely random manner. This 256-bit stream is then hashed in 16-bit segments, each of which is transmitted on a pseudo-random carrier frequency channel. The spectrum is thus spread over a wide frequency band, and data contained within the spectrum simply appear as noise.

If noise is added in the transmission channel, which is usually the case, it will add to the spectrum. On the receiving side where the pseudo-random FH sequence and the pseudo-random seed value of the LFSR are known, noisy data are de-spread, de-interleaved, and descrambled prior to applying reverse FEC (Fig. 5.9).

Since only the emitter and receiver know the spreading, interleaving, and scrambling processes, the noise that has effectively been added in the channel is spread on the receiver side because of the reverse data processing in Wavenis. That is, some consecutive bits destroyed by interferers are spread all over the de-processed received stream.

Reverse FEC is then applied and the spread noise is digitally removed by correcting corrupted bits. This process is similar to DSSS with the notable exception that due to the coordination between emitter and receiver, all transmissions can be performed in a narrow band with improved sensitivity.

Benefits are equivalent to the high-gain process of DSSS data processing but also take advantage of a fast FHSS narrowband receiver in order to achieve a high radio link budget with long-range capability. FHSS technology can also support narrowband mono-channel operation devoted to alarm and security applications in sub-GHz European bands by restricting the hop table to one channel.

This sophisticated combination of techniques makes it possible to take advantage of the benefits of a narrowband channel that features high receiver sensitivity (which is not possible with DSSS)—and therefore a high link budget—while also taking full advantage of spread spectrum with efficient digital noise spreading (similar to DSSS).

Wavenis Wake-Up Preamble

In order to establish a connection, a wake-up sequence is sent prior to the data when communication is initiated. The duration of the wake-up preamble depends on the duty cycle (access time) and whether or not the network is synchronized. The wake-up preamble is composed of a continuous sequence of stuff bits, plus a synchronization bit:

- For non-synchronized networks, i.e., for walk-by/mobile monitoring only, meaning no dedicated root in the network, the wake-up preamble lasts a portion of time (typically 1.28 s) of the duty cycle (standby–receive mode).
- When a network is synchronized, the duration is drastically reduced to 50 ms, only taking into account the maximum clock drift between devices.

Wavenis Carrier Sense for Most Use Cases

Because FHSS is an efficient spread-spectrum technique, Carrier Sense (CS) before transmitting is only optional. When activated, CS is mainly used for point-to-point operation in non-synchronized networks when the application shows a low probability of collision due to low communication traffic. ACK can be activated or deactivated depending on application requirements.

Wavenis CSMA/CA for Mission-Critical Applications

For non-synchronized networks, in applications where reliable operation is critical, all wireless communications must be guaranteed. It is therefore important to avoid simultaneous alarm transmissions that usually result in loss of communications and related data. Target applications include security, alarms, and controlling critical processes. To meet these application requirements, Wavenis has implemented a CSMA/CA mechanism (Carrier Sense Multiple Access—Collision Avoidance). In metering applications, for example, because there is no critical requirement concerning safety, CSMA/CA is not activated. It is, however, operational in most alarm systems. ACK can be activated or deactivated depending on application requirements.

The CSMA/CA principle is described as follows:

- A child listens on the RF channel before transmission.
- if the channel is free, the child sends an RTS (Request to Send) to the parent.
- The parent manages potential conflicts and sends back a CTS (Clear-to-Send).
- Child sends data immediately upon receiving the CTS.

Wavenis CSMA/TDMA for Broadcast-Multicast Feedback Only

Wavenis implements a mixed CSMA/TDMA mechanism (Carrier Sense Multiple Access/Time Division Multiple Access) to manage RF uplinks from children to parents if feedback is requested after a broadcast or multicast message (point-to-multipoint communication mode) in a synchronized network.

After receiving a broadcast message, each child allocates a time slot for transmitting. Because parent and child clocks are aligned, the time slot is calculated using a pseudo-random sequence that depends on its PHY address. Before transmission in the time slot, the child performs Carrier Sense (CS). If the channel is busy, the child does not transmit and a new time slot is calculated according to a pseudo-random sequence. A watchdog allows feedback reception to be ended. This CSMA/TDMA process is optimized to avoid current-consuming collision management and to speed up the feedback process.

Other MAC Protocols

The MAC protocols described previously are aimed at reducing energy consumption. However, the recent increase in the number of applications with real-time requirements has motivated studies and investigations about new QoS-aware MAC protocols. We leave the discussion on these protocols for the section on QoS issues (Section 5.4).

5.2.3 Network Layer

The network layer is the third level in the WSN protocol stack. It responds to service requests from the transport layer and issues service requests to the data link layer. In essence, the network layer is responsible for end-to-end, i.e., source-to-destination, packet delivery, whereas the data link layer is responsible for node-to-node, i.e., hop-to-hop, packet delivery. The network layer provides the functional and procedural means of transferring variable-length data sequences from a source to a destination via one or more networks while maintaining the quality of service requested by the transport layer. The network layer performs network routing, flow control, network segmentation/de-segmentation, and error control functions.

Due to the deployment characteristics of WSNs, multi-hop communication may be a good choice for sensor networks with strict consumption and transmission power level requirements. In a multi-hop network, intermediate nodes have to relay packets from the source to the destination node. Those intermediate nodes have to decide which neighbor to forward to. The construction and maintenance of the routing tables needed for reaching the destination node is the crucial task of a distributed routing protocol. This section discusses some mechanisms for routing and forwarding that can be implemented by WSN routing protocols. These mechanisms take into account whether the packet is identified by a unique node identifier, by a set of such identifiers, or by all nodes in the network.

The network layer of the WSNs is usually designed according to the following principles:

- Energy efficiency is always an important consideration.
- WSNs are mostly data-centric. Sensors do not usually have a unique ID, because the overhead of ID maintenance is high. The data themselves are usually more important than knowing which nodes send data.
- An ideal WSN has attribute-based addressing and location awareness.
- Data aggregation: Depending on the application, this can be useful although the energy needed for data aggregation is sometimes higher than the savings.
- The routing protocol needs to be easily integrated with other networks, e.g., the Internet.
- In some cases, the routing protocol must be QoS-aware; thus having specific mechanisms related to the delay and reliability of the traffic flow.

These design principles serve as a guideline when designing a routing protocol for sensor networks and are further explained to emphasize their importance.

5.2.3.1 Forwarding Types

The multi-hop routing was previously introduced as one of the most appropriate routing methods for WSNs. Whenever a source node has to depend on one or several intermediate nodes to forward its packets to its destination node, the result is a multi-hop network. In such networks, intermediate nodes as well as the source node have to decide which neighbor an incoming packet should be passed to. This act of passing a packet is called forwarding. Several options of how to organize the forwarding processes are available.

The simplest forwarding rule is to flood the network by sending an incoming packet to all neighbors. With this method, the packet is sure to arrive at the destination, if source and destination nodes are connected to the same network. To avoid packets circulating endlessly, i.e., if the destination node is not reachable, the packets usually carry some form of expiration date such as time to live (TTL) or maximum number of hops.

An alternative to forwarding the packet to all neighbors is to forward it to an arbitrary one. This method is called gossiping. With this approach, the packet randomly traverses the network in the hope of eventually finding the destination node. Evidently the packet delay can be quite large. However, flooding and gossiping are two extremes of a design range; there are alternative methods in which the source can send out more than a single packet on a random walk or each node can forward an incoming packet to a subset of its neighbors. For example, packets can be forwarded as determined by a topology-control algorithm, which is equivalent to flooding on a reduced topology. This last option is sometimes called controlled flooding.

While those forwarding rules are simple, their performance in terms of the number of sent packets or packet delay is likely to be poor. With some information about the suitability of a neighbor in the forwarding process, the routing performance can be increased. Neighbor suitability is characterized by assessing the cost of sending a packet to its destination via this particular neighbor. These costs can be measured using various metrics, e.g., the minimum number of hops, the minimum energy wasted, the path reliability, or the minimum delay required to reach the destination via the given neighbor. Each node stores the costs of forwarding to its respective neighbors in routing tables.

A common taxonomy used in WSN arena (Royer and Toh, 1999) classifies the routing protocol as either (1) proactive protocols, which are conservative protocols in which nodes try to keep accurate information in their routing tables, or (2) reactive protocols, which do not attempt to maintain routing tables but only construct them when a packet is prepared to be sent to a destination for which there is no available routing information.

5.2.3.2 Data Centricity and Attribute-Based Addressing

In traditional communication networks, the idea of a communication relationship is normally made up of two entities: one that transmits data and one that receives the data. In WSNs, the application is not interested in the identity of a particular sensor node, but rather in the information sent about its surroundings. This usually occurs when the WSN deployment is redundant. and any event can be sensed by multiple nodes where the application is not aware of the identities of the nodes providing the information. This behavior is called data-centricity. A set of nodes involved in a data-centric address is implicitly defined by the data that each node in this address can contribute.

Besides the separation between entity and identity, the data-centric paradigm also supports a time disconnection in the data requests, since the instant at which the reply will be generated is not specified. This is a useful property for event detection applications. This fact provides a more natural form for an application to express its requirements. In addition, the data-centric paradigm can be used to improve energy efficiency.

There are several possible ways to specify the abstract concept of data-centricity. Each one implies a certain set of interfaces that will be used by an application. The three most important are the following: overlay network and distributed hash tables, publish-subscribe interaction paradigm, and databases.

5.2.3.3 Data Aggregation

Data aggregation is perhaps the simplest in-network processing technique. Assume that a sink is interested in obtaining periodical measurements from every sensor node, but that it is only relevant to check whether the average value

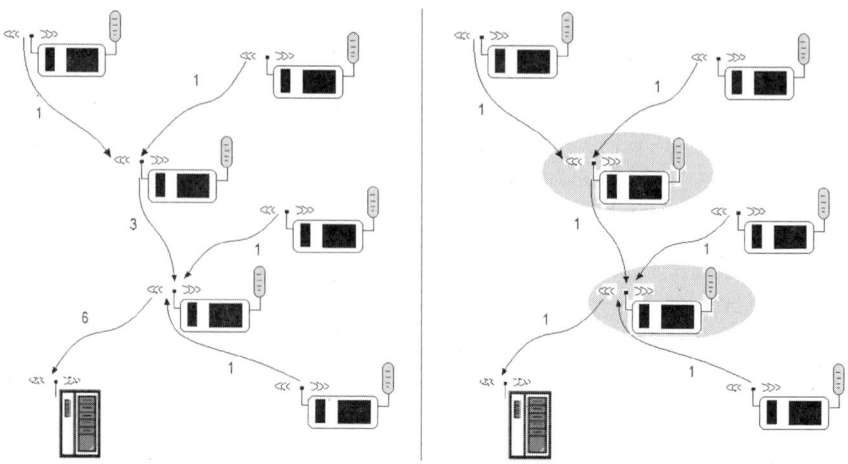

Fig. 5.10 Data aggregation example. (From [Karl and Willig, 2005]. ©John Wiley & Sons Ltd. Reproduced with permission.)

has changed or if the difference between the minimum and maximum values exceeds a certain threshold. In this case, transporting the readings from each sensor node to the sink is not necessary. It is enough to transmit the average value or the maximum and minimum values. Taking into account that the data transmission's energy consumption is normally bigger than the power needed to perform a complex calculation, this solution has large benefits in terms of energy efficiency. The name "aggregation" comes from the fact that the inter-mediate nodes positioned between the source and the sinks aggregate and condense the information, resulting in a new packet.

Figure 5.10 illustrates the aggregation concept. On the left, a few sensor nodes transmit readings to the sink, using multi-hop communication. In total, 13 messages are required; the numbers in the Figure indicate the number of messages that go through a particular link. When the highlighted nodes per-form aggregation, shown on the right, only six messages are needed by means of the average value calculations.

5.2.3.4 Data Delivery Models

Depending on the WSN application, data delivery to the sink can be contin-uous, event-driven, query-driven, or hybrid (Tilak et al., 2002). With contin-uous delivery of data, every node sends data periodically, whereas in the event-driven and query-driven models, data transmissions are activated by events or queries generated by the sink. Some networks apply a hybrid model using a combination of continuous, event-driven, and query-driven delivery. The rout-ing and MAC protocols are highly influenced by the data delivery model since they are directly related to the energy consumption and the path stability. For example, Heinzelman (2000) concludes that for habitat monitoring applications, where data are transmitted continuously to the sink, a hierarchical routing protocol is a more efficient alternative. This is due to the large amount of redundant data this application generates, which can be aggregated in the path to the sink, reducing the traffic and the energy consumed.

5.2.3.5 Study Case: Wavenis Network Capabilities

Wavenis Network Operating Mode in Mobile Monitoring

Networks do not require a synchronization scheme at all for purely mobile network monitoring operations. In this case, handheld terminals, PDAs, or laptops can initiate wireless communication with one or several endpoint devices. When devices are up and running, the default operation mode is as follows:

- Receive–standby mode with a programmable period.
- The typical value is 1.28 s and can range from 12.8 ms to 12.8 s.
- Reception time is only 500 µs if no energy is detected on the RF channel.
- Reception time is just extended to 1.6 ms if energy is detected on the RF channel but without a coherent signal.
- Reception time is extended for normal operation if a useful message is detected.

Wavenis Network Operating Mode in Fixed Monitoring Without Synchronization

For fixed network operations in compliance with the EN300-220 standards, i.e., in Europe and Asia, there is no need to implement a synchronization scheme that consumes more power in wireless networks where low- power consumption is a critical factor. Remember that mobile terminals can also be used for local monitoring in addition to fixed network operations. A fixed network monitoring topology can be

- Centralized monitoring networks with a fixed access point that act as the network root.
- Non-centralized networks—without a root—that meet specific application requirements, such as home comfort and lighting systems.
- When a non-synchronized fixed network is up and running, the default operation mode of wireless nodes is as follows:
  o Receive–standby mode with a programmable period.
  o The typical value is 1.28 s and can range from 12.8 ms to 12.8 s.
  o Reception time is only 500 µs if no energy is detected on the channel.
  o Reception time is just extended to 1.6 ms if energy is detected on the channel but without a coherent signal.
  o Reception time is extended for normal operation if a useful message is detected.

Wavenis Network Operating Mode for Synchronized Fixed Networks

Wavenis fixed networks implement a relaxed synchronization scheme while maximizing ultra low-power features to the greatest extent possible. Fixed networks of this type are generally centralized monitoring networks with a fixed access point that is typically the network root.

In addition to synchronized fixed network operation, mobile handheld terminals, PDAs, and laptops can also perform local (mobile) monitoring services with one or several endpoints. In this case, the mobile device is supposed to be unsynchronized when attempting the first wireless connection.

When a synchronized fixed network is up and running, the default operation mode is as follows:

- Receive–standby mode with a programmable period.
- Typical value is 1.28 s and can range from 12.8 ms to 12.8 s.
- Reception time is only 500 µs if no energy is detected on the channel.
- Reception time is just extended to 1.6 ms if energy is detected on the channel but without a coherent signal.
- Reception time is extended for normal operation if a useful message is detected.
- Parent devices can initiate communication at any time. The "worst-case" access time to child devices is determined by the predefined period (typically 1.28 s).
- Child devices can also initiate communication at any time. The "worst-case" access time to a parent device is determined by the predefined period (typically 1.28 s).
- The receiving carrier frequency changes periodically and hops from one channel to another after each predefined period of time, following a pseudo-random sequence that depends on the MAC address. In addition, for a given parent, all child devices implement a delay that depends on the MAC address and clock of the parent. This means a time shift exists that is a deterministic incorporated delay between the 1.28-s clock of the parent and the 1.28-s clock of all its direct children. Because they "know" each other, the delay is taken into consideration when a radio communication has to be initiated by the parent of a child device.

This mechanism reduces overall wireless network power consumption that is usually the result of overhearing. If all receive time slots were all aligned at the same time on the same channel, the overhearing phenomenon would be maximized.

To simplify the explanation, consider that each parent–child group forms a cluster with its own dedicated delay and its own "listen" channel hop table. To initiate communication (parent–child/child–parent/child–child), the transmitter, which is either the parent or the child, sends a wake-up preamble followed by the data packet. Since clocks are synchronized and channel hop tables and delay are known, the wake-up sequence is short. The wake-up preamble is also

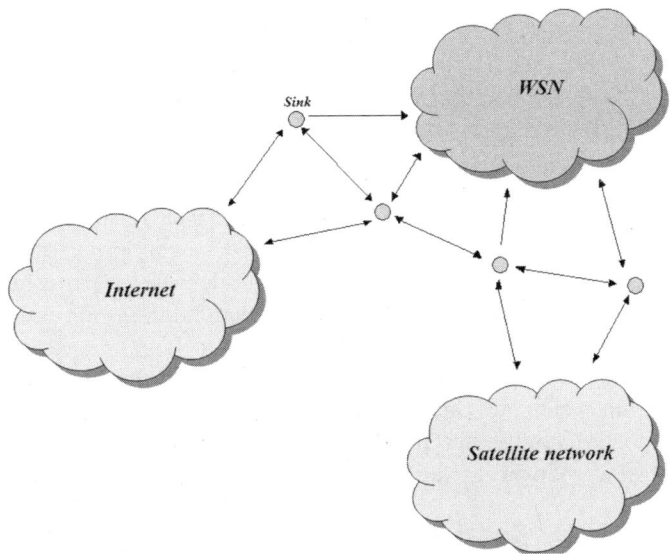

Fig. 5.11 Network interconnection by means of a sink nodes backbone

designed to cover the receive time slot of the target device(s) as well as the maximum drift between parent–child clocks.

5.2.3.6 Network Interconnections

Another key element of the design principles for the network layer is to allow easy integration with other networks such as the satellite network or the Internet. Sinks are the basis of the communication backbone and serve as gateways to other networks (Fig. 5.11). Users may query the sensor networks through the Internet or the satellite network, depending on the purpose of the query or the type of application running.

5.2.4 Transport Layer

The second-highest layer in the WSN protocol stack, the transport layer responds to the services requested from the application layer and issues service requests to the network layer. The transport layer provides dependable data transfers between hosts. It is usually responsible for end-to-end error recovery and flow control and for ensuring complete data transfer.

The purpose of the transport layer is to provide reliable data transfer services between end users, thus relieving the upper layers' responsibility for providing reliable and cost-effective data transfer. The transport layer usually turns the unreliable and very basic service provided by the network layer into a more powerful one. There is a long list of services that can be optionally provided at this level, although none is compulsory. Since not all applications require all services available, some can be wasted overhead or even counterproductive in some cases.

There are some differences between the services provided by the classical network transport protocols and those of WSN. In a classical network such as the Internet, the transport protocol (TCP/UDP) is supposed to transport independent byte streams while intermediate nodes are not informed about end-to-end communication. In a WSN, nodes collaborate and interact with the environment and are aware of the data they carry.

A major requirement for transport protocols is reliability. In sensor networks, reliability refers not only to the eventual delivery of data packets, which is the transport reliability, but also to the ability to detect physical phenomena in the first place. Coverage and deployment of a WSN are thus important considerations.

Several transport protocols for dealing with reliability in a WSN are described in the following sections. These techniques are not exactly referred to as transport protocols, since these are not cleanly placed on top of a network layer protocol. Instead, the unique constraints of sensor networks call for careful cross-layer design.

5.2.4.1 Transport Protocol Objectives

Typically, the major objectives of the transport layer have been

- Network abstraction: The transport layer provides an interface to applications so that the complexities of the data transfer are hidden. Since there is no standard transport protocol in sensor networks, there is no consensus regarding such an interface.
- Reliable data transport: The transport layer must provide data delivery services between the source and the sink with an error control mechanism tailored according to the specific reliability requirement of the application layer.
- Flow control: The receiver of a data stream might temporarily be unable to process incoming packets because of lack of memory or processor power. Flow control has so far not been a research issue in WSNs.
- Congestion control: Congestion occurs when the sources generate more packets than the network can process and the network starts to discard packets. Discarding packets is a waste of energy and an obstacle for achieving reliability or information accuracy. Congestion control mechanisms try either to avoid this situation or to react to it in a reasonable way. One important way of avoiding congestion is to control the rate at which sensor nodes generate packets, i.e., sliding windows.

The transport layer functionalities required to reach those objectives are subject to significant modifications in order to accommodate the particular characteristics of WSNs. Energy, processing, and hardware limitations of sensor nodes further constrain the transport layer protocol design. Many mechanisms implemented by the vastly used transport control protocol (TCP), e.g., retransmission-based error control mechanisms and window-based, additive-increase/multiplicative-decrease congestion control mechanisms, may not be feasible for WSN domains and thus may lead to a waste of scarce resources.

As we stated earlier, it is beneficial to implement reliability, flow control, and other mechanism related to data transfer, not in a single protocol running on top of a network layer, but rather in a combination of several mechanisms working on different layers.

Some of the particular challenges for transport protocols in WSNs include the following:

- WSNs are multi-hop wireless networks with homogeneous/heterogeneous nodes. TCP has several drawbacks when used over wireless channels; thus, a WSN is not an easy environment for TCP.
- Any transport protocol must adapt to the stringent energy constraints, memory constraints or computational constraints of sensor nodes. Significant engineering efforts would be required to run heavyweight protocols like TCP on such nodes.
- Generally, transport protocols do not have good behavior with dynamic topologies.

5.2.4.2 Reliable Data Transport

The problem of reliable transport over wireless multi-hop networks like WSNs is not an easy one to solve. Three main sources of packet losses can be found:

- The wireless channel is inclined to introduce transmission errors. Either transmissions from different nodes can collide or other failures in nodes can produce package losses.
- Packets can be discarded in the network due to congestion, i.e., intermediate nodes' overload.
- The receiver might discard packets because they arrive too quickly, implying a failure in flow control.

Congestion Control

There are two major causes of congestion in WSNs. The first is when the packet arrival rate exceeds the packet service rate. This is more likely to occur at sensor nodes near the sink, since normally they carry more upstream traffic. The second cause relates to performance aspects of the link layer such as contention, interference, and bit error rate.

Congestion in WSNs has a direct impact on energetic efficiency and on QoS parameters. For example, congestion may cause buffer overflow, which could lead to large queuing delays and higher loss rates. Packet loss not only degrades the reliability and QoS of the application, but also wastes node energy. The congestion can also degrade the link utilization. Furthermore, link-level congestion results in transmission collisions in contention-based link protocols such as CSMA. Collisions during transmissions increase packet service time and waste energy, so congestion in WSNs must be efficiently controlled, either to suppress it or to decrease its harmful effects. Typically, there are three mechanisms for controlling congestion: congestion detection, congestion notification, and rate adjustment.

- Congestion detection: In TCP, the congestion is observed or deduced by end nodes based on a timeout or redundant acknowledgments. In WSNs, proactive methods are preferred. A common mechanism would be to concrete a queue length (Hull et al., 2004; Wan et al., 2003), a service time (Wang et al., 2006), or the packet service time ratio over packet interarrival time at the intermediate nodes (Wang et al., 2006). In WSNs with collision-based MAC protocols such as CSMA, the channel load can be measured and used as a congestion indication.
- Congestion notification: After detecting congestion in the network, the transport protocol needs to propagate data about congestion from the congested nodes to the upstream or source nodes that contribute to the congestion. The approach to disseminating congestion data can be classified into implicit congestion notification and explicit congestion notification. Explicit congestion notification uses special control messages to notify the involved nodes that congestion is occurring, by means of suppression messages, see for example (Wan et al., 2003). Implicit congestion notification is included in normal data packets, usually a bit inside the packet.
- Rate adjustment: When receiving a congestion indication, the node can adjust its data transmission rate. If a single congestion notification (CN) bit is used, an additive-increase/multiplicative-decrease (AIMD) scheme or one of its variants can be applied (Wan et al., 2003; Akan and Akyildiz, 2005). If the protocol implements additional information about congestion, more accurate rate adjustment schemes can be adopted (Wang et al., 2006; Ee and Bajcsy, 2004).

Error Recovery

In wireless environments, both congestion and bit error can cause packet losses that degrade the end-to-end reliability and QoS, while packet losses imply a decrease in energy efficiency. Other factors resulting in packet losses include node failures, wrong or outdated routing information, and depletion of energetic resources. There are two general options to recover from packet losses. Increase the source sending rate or a retransmission-based loss recovery. The first approach, which is also used in event-to-sink reliable transport (ESRT) (Akan and Akyildiz, 2005), works well for guaranteeing event reliability in event-driven applications that do not require packet reliability; nevertheless, compared to loss recovery, this method is not energy-efficient. Loss recovery is a more active and energy-efficient method and can be implemented in both the link and transport layers. Loss recovery in the link layer operates hop by hop, whereas loss recovery in the transport layer is usually an end-to-end method. Two approaches for loss recovery that can work well for WSNs consist of loss notification and detection and retransmission recovery.

Loss Notification and Detection

Since loss can be much more common in WSNs than in wired networks, loss detection mechanisms have to be designed more carefully. A typical approach is to include a sequence number in the packet header. The sequence number's continuity can be used to detect packet loss. Loss detection and notification can be end to end or hop by hop. In the end-to-end approach, the endpoints (destination or source) are responsible for the loss detection and notification. In the hop-by-hop method, the intermediate nodes detect and notify the packet loss.

Several reasons make the end-to-end approach not effective for WSNs:

- Control messages used for end-to-end loss detection use a return path consisting of several hops, which is not energy-efficient.
- Control messages travel through multiple hops and can be lost with a high probability due to link-layer errors or congestion.
- End-to-end loss detection inevitably leads to an end-to-end retransmission for the loss recovery. However, end-to-end retransmission wastes more energy than hop-by-hop retransmission.

In a hop-by-hop detection and notification scheme, only two neighboring nodes will be responsible for the loss detection, and they can activate the local retransmission, which is much more energy-efficient than the end-to-end approach. Hop-by-hop loss detection can be categorized as receiver-based or emitter-based, depending on where the loss is detected.

The detection and notification can also identify the reason for the packet loss, which can be used to improve system performance. For example, if packet loss is caused by buffer overflow, source nodes need to decrease their sending rate. However, if channel noise is the cause, then the sending rate does not need to be reduced in order to maintain high link utilization and throughput.

Loss Recovery Based on Retransmission

The retransmission of lost or damaged packets can be also end to end or hop by hop. In the end-to-end approach, the source performs the retransmission, whereas in hop-by-hop retransmission an intermediate node that intercepts the loss notification searches in its local buffer for the lost packet. If it finds a copy of the lost packet, it retransmits the packet; otherwise, it relays loss notification upstream to other intermediate nodes.

5.2.5 *Application Layer*

The application layer is the last level of the WSN protocol stack. It interfaces directly with the application, performs common services for the application processes, and issues requests to the transport layer. The common application layer services provide semantic conversion between associated application processes. The application layer of the five-layer WSN protocol stack corresponds to the application layer, the presentation layer, and the session layer in the seven-layer OSI model.

Although many application areas for sensor networks have been defined and proposed, potential application layer protocols for sensor networks remain a largely unexplored region. Some application protocols for WSNs are the Sensor Management Protocol (SMP), the Task Assignment and Data Advertisement Protocol (TADAP), and the Sensor Query and Data Dissemination Protocol (SQDDP).

The SMP has been proposed for application layers in WSNs, making lower levels transparent. It handles data aggregation, attribute-based naming, clustering, location finding, time synchronization, turning nodes off and on, getting status, reconfiguring, authentication, key distribution, and security.

5.3 Routing in WSN

5.3.1 *Need for New Routing Protocols*

There are several reasons why it is necessary to think about new routing methods in WSNs that have not been used in conventional networks:

- It is not possible to construct a global direction scheme for the deployment of several nodes. Therefore, the classic IP protocols are not applicable in WSNs. Often, it is more important to collect the data than to know the identifiers of the nodes that have sent the data.
- Unlike typical communication networks, almost all WSN applications need to send a flow of continuous information, originating in multiple node regions, to a certain sink. Usually, WSNs are data-centric routing networks where the data are collected with certain attributes taken into account. For example, to obtain data temperatures over 23°C, only the application that is sensing temperatures over 23°C will transmit the data.
- The data traffic generated has considerable redundancy since the neighboring nodes detecting the same event will produce similar data. This

redundancy will have to be used appropriately by the routing protocols to improve both energy consumption and bandwidth use.

- Because the nodes have strongly limited resources such as transmission power, energy, processing capacity, and storage, the resources must be carefully managed.
- In most application scenarios, the WSN nodes are generally stationary in the deployment, except in some cases when mobile nodes are used, thereby producing frequent and unpredictable changes in the network topology.
- Since the data collection is normally based on localization, it is important to always be aware of the sensors' positions. Using GPS hardware for this function is generally not feasible. The methods based on triangulation (Bulusu et al., 2000) are good alternatives, since they allow the nodes to approximately calculate their position depending on the signal power from other known reference points. Bulusu et al. (2000) showed that the algorithms based on triangulation can work very well under conditions where only a few nodes are aware of their position, for example, using GPS hardware. However, the approaches that do not use GPS are more advisable (Intanagonwiwat et al., 2000).

Next we classify suitable routing models for WSNs and the most important protocols described in the WSN literature.

5.3.2 Routing Techniques and Protocols in WSNs

A classification of routing procedures widely accepted in the WSN literature (see Fig. 5.12) divides routing in WSNs into flat-based routing, hierarchical-based routing, and localization-based routing depending on the network structure. Generally, in flat-based routing, the same functionality is assigned to every node. However, in hierarchical-based routing, the nodes play different roles in the network. In localization-based routing, the positional information is used to adequately route the data. A routing protocol will be considered adaptive if it can adapt to the current network conditions and available energy levels. In addition, these protocols can be based on multi-path routing, query, negotiation, or quality of service, among others depending on the protocol functioning.

Finally, the routing protocols can be classified into three categories depending on the method used for finding the path: proactive, reactive, and hybrid. In proactive protocols, every path is computed before needed, whereas reactive protocols compute the paths on demand. The hybrid protocols use a combination of the two protocols.

5.3.2.1 Flat Routing

The first category of routing protocols is the flat routing protocols. In networks using flat routing protocols, every node usually plays the same role where the

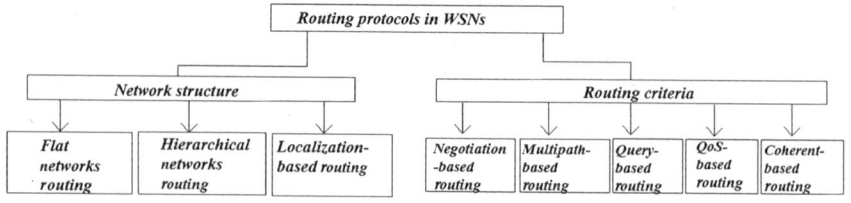

Fig. 5.12 Taxonomy of routing protocols in WSNs

nodes collaborate in the event-sensing task. Due to the large number of nodes, assigning a global identifier to each node is not feasible. Often data-centric networking is carried out, where the sink sends queries to certain regions and waits to receive the information. In data-centric protocols, such as SPIN and directed diffusion (Intanagonwiwat et al., 2000), energy can be saved by data negotiation and redundancy suppression. These protocols have motivated the design of many others based on similar concepts.

Heinzelman et al. (1999) and Kulik et al. (2002) proposed a family of adaptive protocols, called Sensor Protocols for Information via Negotiation (SPIN). These protocols disseminate the information from each node through the network, assuming that every node is a potential base station. This allows a user to request any node and to immediately get the information required by means of the query-driven delivery model. These protocols use the fact that the nearby nodes sense similar information, distributing only the data that other nodes do not have.

The SPIN family uses data negotiation and algorithms for resource adaptation. The nodes running SPIN assign a high-level name to describe the data collected completely (called meta-data). Before beginning data transmission, the nodes perform a meta-data negotiation between each other. This function assures that no redundant data will be traveling through the network. In addition, SPIN can access the current energy levels of each node and thus can adapt the protocol it is running based on the remaining energy.

The SPIN family is designed to solve the deficiencies of classic flooding by data negotiation and resources adaptation. This design is based on two main ideas:

- The nodes should work more efficiently and save more energy by meta-data transmission instead of sending all data.
- The flooding techniques waste energy and bandwidth since the nodes send unnecessary data copies.

Intanagonwiwat et al. (2000) introduced a data aggregation paradigm called directed diffusion. Directed diffusion is a data-centric paradigm in the sense that all node-generated information is designated by attribute pairs. The main idea of the data-centric paradigm is combining data coming from several sources (data aggregation) to reduce redundancy and minimize the number of

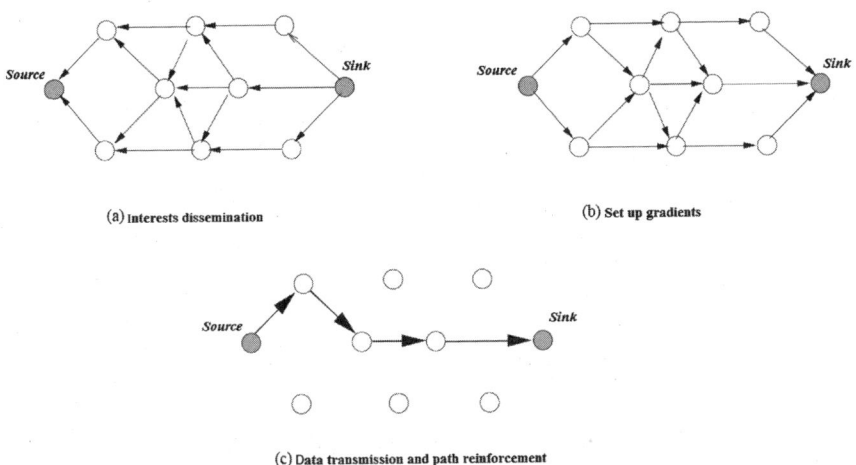

(a) Interests dissemination

(b) Set up gradients

(c) Data transmission and path reinforcement

Fig. 5.13 An example of directed diffusion in WSNs. (From [Al-Karaki and Kamal, 2004].)

transmissions, thus saving energy and prolonging the lifetime. Unlike traditional end-to-end routing, data-centric routing finds paths from multiple sources to a single node that implements the aggregation functions.

In directed diffusion, the sensor nodes sense the events and make information gradients in their respective neighbors. To request data, the sink diffuses some interests, which are disseminated through the network in a multi-hop fashion. The interests describe a task that the network must perform. Each node that receives an interest sets a gradient to the sending node. This process continues until the gradient is established from the source to the sink. A gradient is specified by an attribute value and an address. The gradient intensity can vary depending on the node, originating different information flows. In this phase, loops are not checked, since they are suppressed in later phases. Figure 5.13 shows an example of directed diffusion in a network.

5.3.2.2 Hierarchical Routing

The hierarchical cluster-based routing that was originally proposed for wired networks is made up of a set of techniques related to scalability and communication efficiency. Thus, the hierarchical routing concept is also used to achieve energy efficiency in WSNs. In hierarchical architectures, the nodes with a higher energy load can be used to perform sensing tasks in the proximities of an objective. This means that the cluster formations and special task assignments to cluster heads can specifically contribute to increasing the scalability, lifetime, and energy efficiency of the whole network. Hierarchical routing is an efficient way of reducing energy consumption by means of data aggregation and data fusion, which minimize the number of messages transmitted to the sink. Hierarchical routing is divided into two levels: One is used to select the cluster heads and the other one is used for routing in particular. However, most of the techniques in this category are not based on routing, but rather on who sends or processes/aggregates the information and when.

Heinzelman (2000) introduced a hierarchical clustering algorithm for sensor networks called Low Energy Adaptive Clustering Hierarchy (LEACH). A protocol based on clusters, LEACH handles the distributed information from the clusters. LEACH randomly selects a few nodes and designates them as cluster heads (CH), periodically rotating this function among the nodes to equally distribute the extra energy consumption. The CH nodes compress the information received from the nodes belonging to its cluster and send an aggregated data packet to the sink in order to reduce the amount of information transmitted. LEACH's specification recommends using a MAC protocol based on TDMA/CDMA to reduce the inter- and intracluster collisions.

LEACH's operation period is divided into two phases: the setup phase and the activity phase. In the setup phase, the clusters are organized and the CHs are selected. In the activity phase, the data are transferred to the sink.

Although LEACH is able to increase the network's lifetime, it does have a disadvantage. LEACH assumes that every node can transmit with sufficient power to reach the sink in a single hop and that every node is computationally prepared to house the different MAC protocols required. This fact does not make LEACH very feasible for WSNs deployed in wide zones. However, another system has improved upon LEACH: PEGASIS and "hierarchical PEGASIS" protocols in which the nodes form chains and the multi-hop method is used for routing.

Manjeshwar and Agrawal (2001, 2002) describe two hierarchy routing protocols: TEEN (Threshold-sensitive Energy Efficient sensor Network protocol) and APTEEN (Adaptive Periodic Threshold-sensitive Energy Efficient sensor Network protocol). They proposed these protocols for applications with real-time requirements. In TEEN, the sensors constantly take readings from the environment, but the data are transmitted using a lower frequency. The cluster head will send two values, the triggering value and the minimum threshold value, to the cluster members indicating the value of the measured attribute and the granularity of the reports. The minimum threshold level's goal is to reduce the number of transmissions since the nodes will transmit only when sensing values placed inside an interest range. On the other hand, a triggering level, which provides the granularity of the reports, also contributes to reducing the transmissions, by activating the transmission process only when the sensing value experiences a certain variation. When finer granularity is required, more energy is needed due to the increase in the number of transmissions.

APTEEN is a hybrid protocol that modifies and improves the TEEN protocol in several ways. APTEEN can vary the rate or the threshold values used in the TEEN protocol according to user needs and the type of application. APTEEN uses a TDMA schedule modified to implement the hybrid network. The main characteristic of APTEEN is the combination of proactive and reactive politics, which offers greater flexibility, allowing the user to set the report periods and the threshold values.

Another hierarchy protocol is SMECN (Small Minimum Energy Communication Network) (Rodoplu and Meng, 1999). The goal of SMECN is to divide the network into more energy-efficient subnetworks, i.e., clusters. SMECN identifies a forwarding region for each node. The forwarding region consists of an area where data transmission is more energy-efficient than in other network regions. The main idea of SMECN is to find a subnetwork that has fewer nodes and thus requires fewer hops to reach the sink. Because SMECN is autoconfigurable, it can adapt dynamically to node failures.

5.3.2.3 Location-Based Routing

In location-based routing, sensor nodes are identified by their location. The location is required in order to calculate the distance between two particular nodes so that energy consumption can be estimated. There are three main approaches to determine a node's position (Capkun et al., 2001): using information about a node's neighborhood (proximity-based approaches); exploiting the geometric properties of a given scenario (triangulation and trilateration); and trying to analyze characteristic properties of the node's position compared with premeasured properties (scene analysis). Techniques that use the triangulation approach consist of finding the distance between neighboring nodes by taking three reference points. The node coordinates are obtained by the information exchange between neighbors (Bulusu et al., 2000; Capkun et al., 2001). Alternatively, the node's location can be directly available by satellite communication using low-power GPS receptors (Xu and Saadawi, 2001), which is another triangulation example, although its use is not advisable due to energy consumption. Location information is usually used in efficiently routing data. For instance, if the region to be sensed is known, by using the location of sensor, the query can be diffused only to that particular region, significantly eliminating the number of transmissions. Another energy-saving method is to keep many nodes in a hibernation state when there is no activity. The design problem of localized hibernation period schedulers has been described by Chen et al. (2002) and Xu and Saadawi (2001). Many protocols and methods use location-based routing, some of which are described below.

Yu et al. (2001) discusses a method to disseminate queries, which include geographic attributes, to appropriate regions of a network. This protocol, called Geographic and Energy Aware Routing (GEAR), takes nodes' geographical situation and energy consumption into account when nodes are selected. GEAR disseminates particular "interests" through the network using a method similar to directed diffusion. However, GEAR sends these interests only to a concrete network region and not to the whole network, thus saving more energy than directed diffusion.

In GEAR each node stores information about several parameters, essentially the distance, energy, and density of the areas that the data must cross. With these parameters, an estimation of the costs to reach a destination can be calculated.

Several location-based algorithms are discussed by Stojmenovic and Lin (1999). These protocols implement basic distance-, progress-, and direction-based methods, where the key issues are forward direction and backward direction. A source node or any intermediate node selects one of its neighbors according to certain criteria. The routing methods Stojmenovic and Lin (1999) use are MFR (Most Forward within Radius), GEDIR (Geographic Distance Routing), and DIR (a compass routing method). GEDIR is an algorithm that always moves the data packets to the neighbor closest to the destination. However, the algorithm fails when the packet crosses the same edge twice in a row. In most cases, the MFR method sets up the same path to the destination. In the DIR method, the optimum neighboring node is that with the smallest angular distance from the imaginary line joining the current node with the selected destination. The GEDIR and MFR methods are loop-free, while the DIR method can generate loops unless the traffic flows are memorized or a time-stamp system is implemented.

SPAN (Chen et al., 2002) is another location-based algorithm. This protocol selects several nodes as coordinators, taking their positions into account. The coordinators form a backbone network that can be used to forward messages. A node should become a coordinator if two neighbors cannot communicate with each other either directly or via one or two coordinators. The new and existing coordinators are not necessarily neighbors, making this design less energy-efficient due to the two or three hops distance to the neighbors.

Another example is the GAF protocol (Geographic Adaptive Fidelity), which divides the coverage area forming a grid. The nodes in the same compartment can change their activity states, since all nodes are considered equal. Therefore, this protocol requires a previous redundant deployment.

5.3.2.4 Wavenis Mesh Algorithm

While IEEE 802.15.4 defines two different types of modules, a Reduced Function Device and a Full Function Device/PAN coordinator, Wavenis features only one type of device that is low-cost, is self-routing, and offers ultra low-power and long-range capabilities.

The Wavenis mesh algorithm is designed and optimal for ultra low-power, multi-year operation networks. By nature, most applications require devices that are initially placed in an environment and are not expected to evolve or change frequently. However, absolute installations are rare and networks typically change a few times during their life cycle. This means that an algorithm has to be designed to be able to find a new route without severely impacting the average current consumption of battery-operated devices.

The Wavenis ULP mesh algorithm supports the following:

- Setting up self-organizing networks (Service Discovery Protocol)
- Reconnecting devices to the network and self-healing if a route is broken

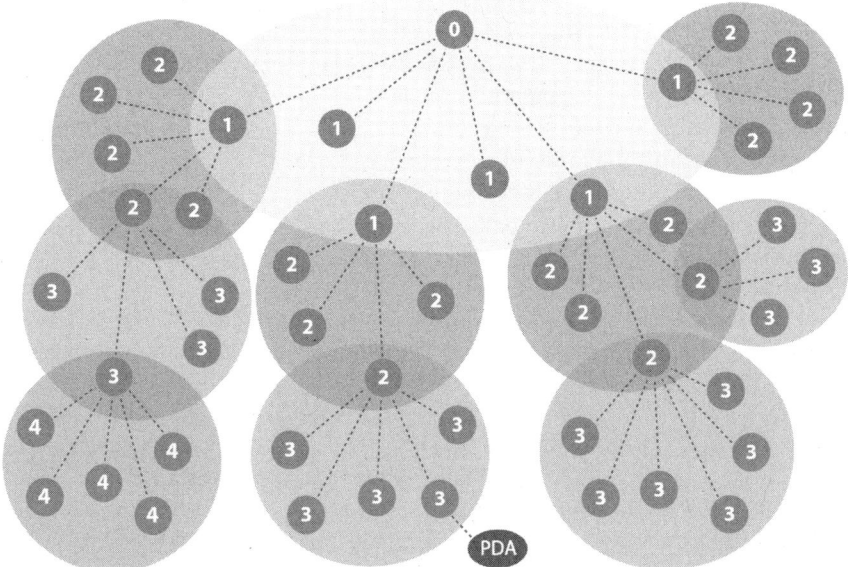

Fig. 5.14 Native Wavenis relay support for deep wireless mesh networks of unlimited size

- Minimizing RF communications while finding the best route by optimally managing device quality of service (QoS)
- Implementing the following
  - Any device that is not yet installed has a predefined "Level 1."
  - A root device (gateway) is "Level 0" (Fig. 5.14).
  - A device connected to a "Level 0" device becomes a "Level 1" device.
  - A device connected to a "Level 1" device becomes a "Level 2" device.
  - A device connected to a "Level 2" device becomes a "Level 3" device.
  - A device connected to a "Level 3" device becomes a "Level 4" device.

The basis of the algorithm is to broadcast requests with selective QoS criteria that allow feedback to be received from only the best possible candidates, thereby avoiding feedback from other weak points. When there is no feedback, the routing step is incremented by changing part of the criteria in order to consider initial conditions that usually have a strong impact on efficiency with energy-saving benefits.

5.3.2.5 Other Routing Protocols

Most protocols for WSN can be classified into one or several of the previous sets. However, due to their particular characteristics, some routing protocols should be classified according to different criteria. These protocols can be classified into multipath-based, query-based, negotiation-based, coherent-based, or QoS-based routing techniques depending on the prótocol operation.

The multipath-based routing protocols use multiple paths to improve network performance, which generally refers to both energy efficiency and fault tolerance. Multipath routing is a typical method for providing communication reliability. The fault tolerance of a protocol can be measured by the probability of an alternative path between the source node and the destination node when the main path fails. The more paths between the source node and the destination node, the higher the fault tolerance. However, this can be possible by means of high energy consumption due to the traffic increase since these alternative paths are kept active by periodic transmissions.

Chang and Tassiulas (2004) proposed an algorithm that routes the data through the path whose nodes have more residual energy, changing the route if one with better characteristics is discovered. The main route is used until the energy of nodes falls below a certain threshold, which occurs when a backup route takes over. Therefore, the network's lifetime can be increased by alternating different paths. Other proposals, such as that by Dulman et al. (2003), use the multipath routing in order to improve the reliability in WSNs. This scheme is useful for transmitting data in environments that are not very reliable. Dulman et al. (2003) use a redundant function that depends on the multipath grade and the fault probability of the available paths. The idea is to divide the original data packet into subpackets to be transmitted through one of the multiple available paths. This function allows the original packet to be reconstructed even if some subpackets are lost.

Nowadays the QoS-based routing protocols are perhaps the most interesting set of protocols from a research point of view. This is mainly due to the introduction of real-time applications, for example, detection and tracking of intruders, environmental monitoring, home and industrial applications, etc. These applications have particular delay and reliability requirements that have posed additional challenges. Transmission of real-time data requires both efficient energy use and QoS awareness routing in order to ensure that accurate measurements are gathered. We have already given an overview of certain protocols; in the following section, we describe them in more detail.

In a WSN application with real-time requirements, the network layer is an essential component to achieve QoS for two main reasons: (1) It is responsible for providing guaranteed paths to join two points and energy efficiency along with reliability; (2) it is used as an intermediary between the MAC and application layers to exchange performance parameters.

Due to the intensive use of resources inherent in real-time applications and the low resource availability in WSNs, the function of the routing protocol is quite difficult. The nature of the environment varies over time; thus, guaranteeing real time is fairly complicated. However, the network can ensure soft real-time or soft QoS (Veres et al., 2001) guarantees. Many mechanisms have been designed to solve the problems produced by the WSN's changeable nature. One of the most extended in the field of WSNs is multi-hop routing. However, other mechanisms are equally or more interesting, as we will explain next.

5.4 WSN Performance: Quality of Service

5.4.1 An Increasing Interest in QoS for WSNs

Research on the network and link layers of the WSN protocol stack has had as its main motivations the architecture and protocol design, energy conservation, and location. Although only a few studies concerning QoS in WSNs have been performed, there are several very interesting works regarding QoS. These papers proposed various protocols and mechanisms for both the MAC and network layers, and almost all have been developed and tested with a simulator. All these approaches supporting QoS in WSNs form a solid base for future work and research in this direction.

An exact definition of QoS in the WSN context is provided in the following section. The most important protocols and mechanisms in order to provide QoS in WSNs that have been proposed in recent papers for both the network and MAC layers are also explained below.

5.4.2 Quality of Service in the WSN Context

In the field of WSNs, the quality-of-service concept refers to the capacity of a network to deliver data both reliably and timely. Generally, a large amount of resources such as high throughput or transport capacity is not enough to satisfy an application's delay requirements. Consequently, the speed with which to propagate information could be as important as the throughput speed. Therefore, the quality-of-service (QoS) guarantee in WSNs in addition to other design network issues such as energy efficiency is an important issue. Previous papers related to QoS in WSNs mostly focused on delay (Lu et al., 1999; Ju and Li, 1999; Luo et al., 2000). Recently, a new issue that attempts to unite two seemingly opposite concepts has emerged: energy efficiency and QoS. This new challenge has motivated several research papers (e.g., Akkaya and Younis, 2003). QoS can be defined by the (R, Pe, D) triplet, where R denotes throughput; Pe refers to reliability measured by the bit error probability or packet loss probability, for example; and D represents delay. For a given R, the reliability of a connection as a function of the delay will follow the general curve shown in Fig. 5.15.

Because QoS is affected by design choices at the physical, medium access, and network layers, an integrated approach to managing QoS is necessary.

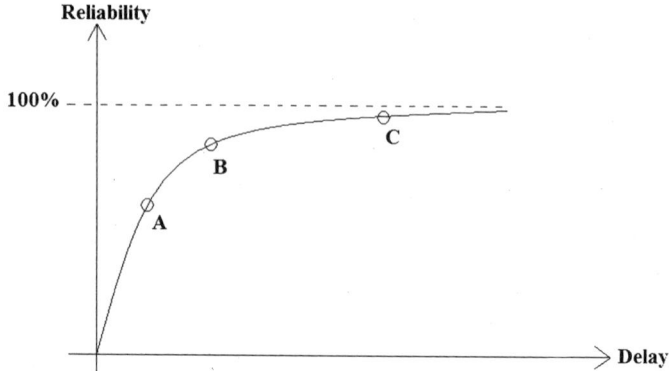

Fig. 5.15 Reliability as a function of the delay. The circles indicate the QoS requirements of different possible traffic classes. (From [Haenggi, 2006].)

5.4.3 QoS MAC Protocols

5.4.3.1 B-MAC

B-MAC (Polastre et al., 2004) stands out for its design and implementation simplicity, which has an immediate effect on memory size occupation and power savings. B-MAC's high-quality design compensates for the fact that it does not implement a specific QoS mechanism. Some parts of this design are specifically intended to improve efficiency in order to avoid collisions, channel occupation efficiency at low and high data rates, tolerance to changeable environments, or scalability for a high number of nodes. Although B-MAC was initially planned for monitoring applications, it can be used in other applications such as target tracking, localization, triggered events, and multi-hop routing. B-MAC also has a high degree of configurability. Keeping all these characteristics in mind, it is clear that B-MAC is a good alternative for applications based on event-driven data delivery models with minimum delay requirements.

5.4.3.2 Z-MAC

ZMAC (Zebra MAC) (Rhee et al., 2005) is a hybrid scheme that combines the advantages of CSMA and TDMA while removing their weaknesses. ZMAC is characterized by an initial period in which a wide time-slot scheduling is performed using a DRAND, a very efficient distributed scheduling algorithm. The initial assignment of slots incurs a high overhead, but the overhead is then spread out over a long network operational period and eventually is compensated for with improved power saving and throughput. ZMAC also implements contention control, thereby avoiding congestion situations. As a result, ZMAC has similar behavior to CSMA under low contention. Under high contention, its behavior is similar to TDMA. This approach is robust enough for dynamic topology changes. These two characteristics are very important for applications with delay and/or reliability requirements.

5.4.3.3 i-GAME

The MAC protocol of the IEEE 802.15.4 standard implements a mechanism called Guarantee Time Slot (GTS). GTS tries to assign an additional time slot for applications with delay requirements but is less efficient in WSNs with a large number of nodes. In order to correct this deficiency, Koubaa et al. (2006) proposed an implicit GTS allocation mechanism (i-GAME). The main idea of i-GAME consists of sharing the same GTS between multiple nodes instead of being exclusively dedicated to a single node. GTS resources are assigned based on an admission control algorithm. This algorithm admits a request if its requirements do not exceed the available resources.

5.4.3.4 MAC for Linear WSN

Watteyne and Auge-Blum (2005) proposed a hard real-time MAC protocol for a network with low-cost sensors (e.g., only one frequency), deployed randomly, with no differentiated nodes (e.g., no router nodes), and unsynchronized (without a global clock). This protocol was planned for a linear network and therefore has no routing considerations. A sink node is situated at one end, where it receives all events originated in the network. This protocol alternates between two operational modes: protected and unprotected. When the network is in the unprotected mode, the transmission speed is near optimal, but collisions could be found. However, in the protected mode, the transmission speed is slower, but the frames are reliably transmitted since the network is collision-free. This characteristic can be interesting for real-time applications with critical requirements.

5.4.3.5 Comparing QoS-Aware MAC Protocols for WSNs

Table 5.1 Compares the various MAC protocols in WSNs in terms of quality of service.

Table 5.1 Comparative Table of MAC Protocols in Wireless Sensor Networks

Protocols	Data Aggregation/ Merging	Scalability	Priority Mechanism	Energy-Aware	Contention-Based
B-MAC	No	High	No	Yes	Yes
Z-MAC	No	High	Yes	Yes	Hybrid
Watteyne and Auge-Blum (2005)	No	Low	Yes	No	Yes
MAC 802.15.4 with i-GAME	No	Medium	Yes	Yes	No

5.4.4 QoS Network Protocols

5.4.4.1 Directed Diffusion

Directed diffusion (Intanagonwiwat et al., 2000) is a data-centric and application-aware paradigm since all data generated by sensor nodes are named by attribute-value pairs. Directed diffusion, unlike traditional end-to-end routing, tries to find routes from multiple sources to a single destination, allowing redundant data aggregation.

The objective of the directed diffusion paradigm is to aggregate the data coming from different sources by deleting redundancy, which drastically reduces the number of transmissions. This has two main consequences: First, the network saves energy and extends its life; second, it counts on a higher bandwidth in the links near the sink node. The latter factor could be quite persuasive in deciding to provide QoS in real-time applications.

The directed diffusion paradigm is based on a query-driven model, which means that the sink node requests data by broadcasting interests. Requests can originate from humans or systems and are defined as pair values, which describe a task to be done by the network. The interests are then disseminated through the network. This dissemination sets up gradients to create data that will satisfy queries to the requesting node. When the events begin to appear, they start to flow toward the originators of interest along multiple paths. This behavior provides reliability for data transmissions in the network.

Another feature of directed diffusion is that it caches network data, generally the attribute-value pair's interests. Caching can increase efficiency, robustness, and the scalability of coordination between sensor nodes, which is the essence of the directed diffusion paradigm.

5.4.4.2 SPIN

This family of data-centric protocols has been discussed in a previous section. The data negotiation method is emphasized ahead due to its usefulness in QoS issues. Nodes running the SPIN protocol assign a high-level name to describe the data they have collected (called meta-data), and meta-data negotiations are performed before any data are transmitted. The main goal of this mechanism is similar to typical aggregation systems. However, this mechanism has an advantage with respect to other systems: It avoids redundant data transmissions for later processing. The network thus increases its life and the bandwidth available, and the nodes are free from the processing load that data aggregation entails.

5.4.4.3 TEEN and APTEEN

TEEN and APTEEN, proposed by Al-Karaki and Kamal (2004) and Manjeshwar and Agrawal (2001), were defined for time-critical applications. These protocols are designed to work even in the event that an abrupt change takes place in the attribute values being measured by the sensors. APTEEN (Adaptive TEEN) is a modification of TEEN that additionally considers the case of periodic measurement transmissions toward the sink node. This protocol implements a very complex query system that makes it possible to achieve three types of queries: historical, one-time, and persistent. All of these queries are carried out by external users through the sink node. The historical and persistent queries do not need QoS requirements, but the one-time queries become critical data with respect to time. In this case, the end user should be aware of his or her geographical position with minimum delay. In order to achieve minimum delay, the system executes a special time-slot management assigned to each node by a TDMA schedule. Furthermore, APTEEN carries out the important task of data aggregation, which is the equivalent of having free bandwidth and energy savings.

5.4.4.4 SAR

Sequential Assignment Routing (SAR) proposed by Sohrabi et al. (2000) was one of the first protocols for WSN that considered QoS issues for routing decisions. SAR makes a routing decision based on three factors: energy resources, QoS planned for each path, and the packet's traffic type, which is implemented by a priority mechanism. To resolve reliability problems, SAR uses two systems consisting of a multipath approach and localized path restoration done by communicating with neighboring nodes. The multipath tree is defined by avoiding nodes with low-energy or QoS guarantees while taking into account that the root tree is located in the source node and its ends in the sink nodes set. In other words, SAR creates a multipath table whose main objective is to obtain energy efficiency and fault tolerance. Although this ensures fault tolerance and easy recovery, the protocol suffers certain overhead when tables and node states must be maintained or refreshed. This problem increases especially when there are a large number of nodes.

5.4.4.5 SPEED

SPEED (He et al., 2003) is another QoS routing protocol for WSNs that provides light real-time end-to-end guarantees. SPEED's QoS mechanism is based on estimation procedures. The application in a node estimates the required speed for a certain delay while taking into account its distance to the sink node. The network layer admits the packet depending on the required speed. Moreover, SPEED is able to recover if the network becomes congested.

SPEED's routing module is called Stateless Non-deterministic Geographic Forwarding algorithm (SNGF). This module implements a distributed database where a node can be selected in order to reach the speed requirement.

5.4.4.6 MMSPEED

MMSPEED (Multipath and Multi-SPEED Routing Protocol) (Felemban et al., 2005) is an innovative packet delivery mechanism for QoS provisioning and focuses on timeliness and reliability. Thus, traffic flow is handled by a combination of service options based on their reliability and timeliness requirements. The method MMSPEED uses to obtain reliability is the typical multi-path routing with a number of paths that depend on the required degree of reliability for the various flows of traffic. In addition, to obtain timeliness, MMSPEED uses a dynamic system that guarantees the packet delivery speed. MMSPEED uses localized geographic forwarding by using only local-node neighbor information. The local decisions imply an inaccuracy problem, which is solved by dynamic compensation, thus fulfilling traffic flow requirements. The intermediate nodes can increase the transmission packet speed if they estimate that the packet cannot fulfill its delay deadline at the current speed.

To give functionality to the QoS mechanisms implemented by MMSPEED, a MAC protocol with a prioritization mechanism should be established. In this sense, the MMSPEED specification recommends the use of 802.11e at the MAC layer with its inherent prioritization mechanism based on the differentiated inter-frame spacing (DIFS). Each speed value is mapped onto a MAC-layer priority class.

The MMSPEED protocol solves many QoS issues related to real-time traffic in WSNs. However, many other aspects such as network-layer aggregation or handling the energy-delay trade-off still need to be dealt with in order to achieve a higher performance level in a deployed WSN.

5.4.4.7 Energy-Aware QoS Routing

Akkaya and Younis (2003) proposed a QoS-aware protocol for real-time traffic generated by a WSN consisting of image sensors. Their protocol implements a priority system that divides the traffic flows in two classes: best-effort and real-time. All nodes use two queues, one for each class of traffic, allowing different kinds of services to be provided. In addition, the protocol implements a multipath-based routing mechanism, which uses an extended version of Dijkstra's algorithm, that can provide certain reliability in data transmissions. The source node chooses a route in order to achieve the end-to-end requirements and then forwards the packet to the next hop neighbor in the route. Each intermediate node classifies the received packet as real-time or best-effort. The scheduling algorithm prevents the best-effort traffic from reducing resources to the real-time traffic. The main disadvantage of this protocol is that it supports only one real-time traffic priority. This characteristic can be appropriate for a network with a single application, but in a network with multiple applications, there could be several types of real-time traffic with different priorities.

5.4.4.8 Comparing QoS–Aware Routing Protocols for WSNs

Table 5.2 compares the various routing protocols for WSNs that take QoS into consideration.

Table 5.2 Comparative Table of Routing Protocols in Wireless Sensor Networks

	Network Topology	Data Delivery Models	Data Aggregation/ Fusion	Traffic Guarantees	Classes of Traffic	Networks Dynamics	Resources Reservation	Scalability
Directed diffusion	Flat	Query-driven and event-driven	Yes	Reliability	Yes	Limited	Yes	Medium
SPIN	Flat	Query-driven and event-driven	Yes (by means of data negotiation)	No	No	Limited	No	Low
TEEN & APTEEN	Hierarchical	Query-driven, event-driven, and continuous	Yes	Certain guarantees of real time	Yes	Fixed sink	No	High
SAR	Flat	Query-driven and event-driven	Yes	Real time and reliability	Yes	No	Yes	Low
SPEED	Flat	Query-driven and event-driven	No	Soft real time	Yes	No	Yes	Low
MMSPEED	Flat	Event-driven and continuous	No	Reliability and real time	Yes	Limited	No	High
Akkaya and Younis (2003)	Hierarchical	Event-driven and continuous	No	Reliability and real time	Yes	Fixed sink	No	Low

5.5 Open Issues in Network and Deployment Technologies

Open issues in network and deployment technologies include the following:

- Quality of Service (QoS) in WSNs. Energy efficiency is a major issue of concern for WSN designers, and various energy-awareness approaches can be found. Although the performance of these protocols and mechanisms is promising in terms of energy efficiency, other factors have not been taken into account. Recently, new applications for WSNs have begun to be commercialized; many of them can be classified as real-time applications. These types of applications have very rigid quality-of-service requirements that usually include delay and reliability. Currently, only a few research studies have been carried out regarding QoS requirements in environments with energy constraints such as sensor networks. Energy-aware QoS routing and MAC protocols are still open issues. Therefore, future protocols for QoS scenarios could balance energy-efficiency and optimization criteria such as latency, reliable data delivery, and compliance with real-time constraints (Martínez et al., 2007a, b, 2008).
- Mobility. Most of the current protocols assume that sensor nodes and sinks are stationary, making mobility an interesting property that could be included in WSN protocols. In fact, current applications could require this characteristic, such as real-time target tracking in battle situations. Mobile nodes can be used to avoid holes in the coverage and to generate information to be transmitted through the network. If this information is not properly handled, energy can be wasted. New routing algorithms are needed to make sure there is an overlapping of coverage in the event of mobility and topology changes in energy-constrained environments such as the WSNs.
- Localization. Another important issue is the automatic localization of wireless sensor nodes. Substantial theory and systems research has been done on sensor network localization; studies have shown that having GPS hardware on every sensor node can be very costly. A number of localization methods have been proposed, yet there are significant challenges to ensuring that localization is self-configuring and robust not only in laboratory settings but also in unknown environments where real-world applications will be deployed. The beacon-based localization proposals could be the basis of a possible solution to that problem.
- Cross-layer design. A typical transport protocol may not be feasible in WSNs due to its retransmission overhead, e.g., TCP. On the contrary, a cross-layer approach could be the solution to provide reliability, flow control, and other mechanisms commonly implemented in transport protocols. Cross-layer optimization is very important in order to achieve QoS, where the application requirements, specified at the application layer, must be efficiently and correctly mapped onto the dependent performance parameters of the network and data link (MAC) layers. Despite notable research efforts (Goldsmith and Wicker, 2002; van Der Schaar and Sai Shankar,

2005), mapping the QoS parameters across different layers for cross-layer optimization still can be considered an open research issue.

- Adaptability to network objectives. The amount of energy spent on a particular task depends on how critical the current application objectives are. For example, the same WSN used for a low-frequency, continuous monitoring application may later be in a mission-critical tracking or emergency threat alert application. In another example, when the goals are more critical, energy-savings requirements become secondary with respect to latency and throughput. Therefore, one goal is to develop medium access plans that are able to change their mission or objective when necessary. Although several attempts have been made, more work must be done in this area. Another issue that should be exhaustively investigated is traffic patterns. WSN applications exhibit a few specific traffic patterns; much could be gained by studying these patterns better in the MAC design, instead of supporting arbitrary communication patterns.

- Over-the-air-programming of WSN nodes. The nodes' software for a deployed WSN may need to be reprogrammed or updated from time to time. This task must be carried out reliably, but only a few networking protocols were designed to reliably carry out such a task in a multi-hop network. The main difficulty is acknowledging the packets in a joint multi-hop/multicast communication.

- Scalability. Most WSN applications require a larger node count as well as a higher density than other types of networks. This should be taken into consideration when designing medium access control and routing protocols. Less scalable protocols may lead to unbearable overheads, which may cause unnecessary energy consumption in certain nodes or even lead to network failure. There also may be severe QoS reductions; in this context, the scalability of currently available MAC and routing protocols should be further investigated.

- Synchronization. In a distributed system such as a WSN, time synchronization is an important issue for several reasons. First, the coordination and collaboration of sensor nodes needs a common timescale. Second, the nodes must coordinate their active and idle modes, which, of course, requires synchronization. Finally, some scheduling algorithms such as TDMA need synchronization. Nowadays, the most precise solutions are those that use at least one node equipped with a GPS receiver; they may use more than one node depending on the scale of the WSN (He and Kuo, 2006). Because GPS hardware increases the total cost of a WSN significantly, one of the challenges is obtaining a precise time synchronization algorithm without the need for a GPS receiver.

- Node deployment. In most WSN designs, sensor nodes are randomly or uniformly distributed because of their simplicity. However, node deployment greatly impacts wireless systems and can affect both energy efficiency and QoS. Several studies have been done on the possibility of a non-uniform, power-aware distribution plan (Liu et al., 2006). Therefore, node deployment methodologies remain an important open issue.

- Security. Just as classical IP networks are exposed to attacks compromising their security, the WSNs also are at risk. The attacks usually proceed from malicious nodes. A malicious node can misrepresent its identity in the network and issue route error messages to misdirect the path or drop incoming packets, among other possible attacks. If the network is intrusion-tolerant, then a malicious node can only compromise a very small number of nodes in its vicinity, rather than causing widespread damage in sensor networks. Several interesting approaches about security in WSNs have been proposed by Agah et al. (2006), Deng et al. (2006), and Khalil et al. (2007).

References

Agah A,Basu K,Das SK (2006)Security enforcement in wireless sensor networks: A framework based on non-cooperative games.*Pervas Mobile Comp*2(2):137–58.

Akan ÖB,Akyildiz IF (2005)Event-to-sink reliable transport in wireless sensor networks. *IEEE/ACM Trans Netw* 13(5):1003–16.

Akkaya K, Younis M (2003) An energy-aware QoS routing protocol for wireless sensor networks. In*Proceedings of the 23rd International Conference on Distributed Computing Systems Workshops*, pp. 710–715.

Akyildiz IF, Su W, Sankarasubramaniam Y, et al. (2002)A survey on sensor networks. *IEEE Commun Mag* 40(8):102–14.

Al-Karaki JN, Kamal AE (2004) A taxonomy of routing techniques in wireless sensor networks. In Ilyas M, Mahgoub I (Eds) *Handbook of Sensor Networks: Compact Wireless and Wired Sensing Systems*, CRC Press, Boca Raton, FL.

Buettner M, Yee G, Anderson E, et al. (2006) X-MAC: A short preamble MAC protocol for duty-cycled wireless sensor networks. University of Colorado at Boulder, Technical Report CU-CS-1008-06.

Bulusu N, Heidemann J, Estrin D (2000) GPS-less low-cost outdoor localization for very small devices. *IEEE Pers Commun* 7(5):28–34.

Capkun S, Hamdi M, Hubaux JP (2001) GPS-free positioning in mobile ad-hoc networks. In *Proceedings of the 34rd Annual Hawaii International Conference on System Sciences*.

Chang JH, Tassiulas L (2004) Maximum lifetime routing in wireless sensor networks. *IEEE/ACM Trans Netw* 12(4):609–19.

Chen JC, Sivalingam KM, Agrawal P, et al. (1998) A comparison of MAC protocols for wireless local networks based on battery power consumption. In *Proceedings of the 17th Annual Joint Conference of the IEEE Computer and Communications Societies(INFOCOM '98)*,Vol. 1, pp. 150–157.

Chen B, Jamieson K, Balakrishnan H, et al. (2002) SPAN: An energy-efficient coordination algorithm for topology maintenance in ad hoc wireless networks. *Wirel Netw* 8(5):481–94

Deng J, Han R, Mishra S (2006) INSENS: Intrusion-tolerant routing for wireless sensor networks. *Comp Commun* 29(2):216–30.

Dulman S, Nieberg T, Wu J, et al. (2003) Trade-off between traffic overhead and reliability in multipath routing for wireless sensor networks. In *Proceedings of the 2003 IEEE Wireless Communications and Networking Conference (WCNC 2003)*, Vol. 3, pp. 1918–1922.

Ee CT, Bajcsy R (2004) Congestion control and fairness for many-to-one routing in sensor networks. In *Proceedings of the 2nd International Conference on Embedded Networked Sensor Systems (SenSys '04)*. pp. 148–161.

Felemban E, Lee CG, Ekici E, et al. (2005) Probabilistic QoS guarantee in reliability and timeliness domains in wireless sensor networks. In *Proceedings of the 24th Annual Joint*

Conference of the IEEE Computer and Communications Societies (INFOCOM2005), Vol. 4, pp. 2646–2657.

Goldsmith AJ, Wicker SB (2002) Design challenges for energy-constrained ad hoc wireless networks. *IEEE Wirel Commun* 9(4):8–27.

Haenggi M (2006) Opportunities and challenges in wireless sensor networks. In *Smart Dust: Sensor Network Applications, Architecture and Design*, Taylor & Francis, Boca Raton, FL.

He L, Kuo GS (2006) A novel time synchronization scheme in wireless sensor networks. In *Proceedings of the 63rd IEEE Vehicular Technology Conference (VTC 2006-Spring)*, pp. 568–572.

He T, Stankovic JA, Lu C, et al. (2003) SPEED: A stateless protocol for real-time communication in sensor networks. In *Proceedings of the 23rd International Conference on Distributed Computing Systems*, pp. 46–55.

Heinzelman WB (2000) Application-specific protocol architectures for wireless networks. PhD thesis, Massachusetts Institute of Technology, Cambridge, MA.

Heinzelman WR, Kulik J, Balakrishnan H (1999) Adaptive protocols for information dissemination in wireless sensor networks. In *Proceedings of the 5th Annual ACM/IEEE International Conference on Mobile Computing and Networking (MobiCom '99)*, pp. 174–185.

Heinzelman WR, Chandrakasan A, Balakrishnan H (2000) Energy-efficient communication protocol for wireless microsensor networks. In *Proceedings of the 33rd Annual Hawaii International Conference on System Sciences*.

Hightower J, Borriello G (2001) A survey and taxonomy of location systems for ubiquitous computing. University of Washington, Technical Report UW-CSE 01-08-03.

Hull B, Jamieson K, Balakrishnan H (2004) Mitigating congestion in wireless sensor networks. In *Proceedings of the 2nd International Conference on Embedded Networked Sensor Systems (SenSys '04)*, pp. 134–147.

Intanagonwiwat C, Govindan R, Estrin D (2000) Directed diffusion: A scalable and robust communication paradigm for sensor networks. In *Proceedings of the 6th Annual International Conference on Mobile Computing and Networking (MobiCom'00)*, pp. 56–67.

Ju JH, Li VOK (1999) TDMA scheduling design of multihop packet radio networks based on Latin squares. In *Proceedings of the 18th Annual Joint Conference of the IEEE Computer and Communications Societies (INFOCOM '99)*, Vol. 1, pp. 187–193.

Karl H, Willig A (2005) *Protocols and Architectures for Wireless Sensor Networks*. John Wiley & Sons Ltd.,New York.

Khalil I, Bagchi S, Shroff N (2007) Analysis and evaluation of Secos, a protocol for energy efficient and secure communication in sensor networks. *Ad Hoc Netw* 5(3):360–91.

Koubaa A, Alves M, Tovar E (2006) *Proceedings of the 18th Euromicro Conference on Real-Time Systems*, pp. 183–192.

Kulik J, Heinzelman W, Balakrishnan H (2002) Negotiation-based protocols for disseminating information in wireless sensor networks. *Wirel Netw* 8(2/3):169–85.

Liu Y, Ngan H, Ni LM (2006) Power-aware node deployment in wireless sensor networks. In *Proceedings of the 2006 IEEE International Conference on Sensor Networks, Ubiquitous, andTrustworthy Computing*, pp. 128–135.

Lu S, Bharghavan V, Srikant R (1999) Fair scheduling in wireless packet networks. *IEEE/ACM Trans Netw* 7(4):473–89.

Luo H, Lu S, Bharghavan V (2000) A new model for packet scheduling in multihop wireless networks. In *Proceedings of the 6th Annual International Conference on Mobile Computing and Networking (MobiCom '00)*, pp. 76–86.

Manjeshwar A, Agrawal DP (2001) TEEN: A routing protocol for enhanced efficiency in wireless sensor networks. In *Proceedings of the 15th International Parallel and Distributed Processing Symposium*, pp. 2009–2015.

Manjeshwar A, Agrawal DP (2002) APTEEN: A hybrid protocol for efficient routing and comprehensive information retrieval in wireless sensor networks. In *Proceedings of the International Parallel and Distributed Processing Symposium*. pp. 195–202.

Martínez JF, García AB, Corredor I, et al. (2007a) Trade-off between performance and energy consumption in wireless sensor networks. *Lect Notes Comp Sci* 4725:264–71.

Martínez JF, García AB, Corredor I, et al. (2007b) Modelling QoS for wireless sensor networks. *IFIP* 248:143–54.

Martínez JF, García AB, Corredor I, et al. (2008) Guaranteeing QoS in wireless sensor networks. In *Wireless Quality-of-Service: Techniques, Standards and Applications*, Auerbach, US.

Polastre J, Hill J, Culler D (2004) Versatile low power media access for wireless sensor networks. In *Proceedings of the 2nd International Conference on Embedded Networked Sensor Systems (SenSys '04)*, pp. 95–107.

Pottie G, Kaiser W (2005) *Principles of Embedded Networked SystemsDesign*. Cambridge University Press, New York.

Rajendran V, Obraczka K, Garcia-Luna-Aceves JJ (2003) Energy-efficient collision-free medium access control for wireless sensor networks. In *Proceedings of the 1st International Conference on Embedded Networked Sensor Systems*, pp. 181–192.

Rhee I, Warrier A, Aia M, et al. (2005) Z-MAC: A hybrid MAC for wireless sensor networks. In *Proceedings of the 3rd International Conference on Embedded Networked Sensor Systems (SenSys '05)*, pp. 90–101.

Rodoplu V, Meng TH (1999) Minimum energy mobile wireless networks. *IEEE J Sel Areas Commun* 17(8):1333–44.

Royer EM, Toh CK (1999) A review of current routing protocols for ad hoc mobile wireless networks. *IEEE Pers Commun* 6(2):46–55.

Sohrabi K, Pottie GJ (1999) Performance of a novel self-organization protocol for wireless ad-hoc sensor networks. In *Proceedings of the IEEE 50th Vehicular Technology Conference (VTC 1999)*, Vol. 2, pp. 1222–1226.

Sohrabi K, Gao J, Ailawadhi V, et al. (2000) Protocols for self-organization of a wireless sensor network. IEEE Pers Commun 7(5):16–27.

Stojmenovic I, Lin X (1999) GEDIR: Loop-free location-based routing in wireless networks. In *Proceedings of IASTED—International Conference on Parallel and Distributed Computing and Systems*, pp. 1025–1028.

Tilak S, Abu-Ghazaleh NB, Heinzelman W (2002)A taxonomy of wireless micro-sensor network models. *SIGMOBILE Mobile Comp Commun Rev* 6(2):28–36.

van Der Schaar M, Sai Shankar N (2005) Cross-layer wireless multimedia transmission: Challenges, principles, and new paradigms. *IEEE WirelCommun* 12(4):50–8.

Veres A, Campbell AT, Barry M, et al. (2001) Supporting service differentiation in wireless packet networks using distributed control. *IEEE J Sel Areas Commun* 19(10):2081–93.

Wan CY, Eisenman SB, Campbell AT (2003) CODA: Congestion detection and avoidance in sensor networks. In*Proceedings of the 1st International Conference on Embedded Networked Sensor Systems (SenSys '03)*, pp. 266–279.

Wang C, Sohraby K, Lawrence V, et al. (2006) Priority-based congestion control in wireless sensor networks. In*Proceedings of the 2006 IEEE International Conference on Sensor Networks, Ubiquitous, and Trustworthy Computing*, pp. 22–31.

Watteyne T, Auge-Blum I (2005) Proposition of a hard real-time MAC protocol for wireless sensor networks. In*Proceedings of the 13th IEEE International Symposium on Modeling, Analysis, and Simulation of Computer and TelecommunicationSystems*, pp. 533–536.

Woesner H, Ebert JP, Schlager M, et al. (1998) Power-saving mechanisms in emerging standards for wireless LANs: The MAC level perspective. *IEEE Pers Commun* 5(3):40–8.

Wong KD (2004) Physical layer considerations for wireless sensor networks. In *Proceedings of the 2004 IEEE International Conference on Networking, Sensing and Control*,Vol. 2, pp. 1201–1206.

Xu S, Saadawi T (2001) Does the IEEE 802.11 MAC protocol work well in multihop wireless ad hoc networks? *IEEE Commun Mag* 39(6):130–7.

Ye W, Heidemann J, Estrin D (2002) An energy-efficient MAC protocol for wireless sensor networks. In *Proceedings of the 21st Annual Joint Conference of the IEEE Computer and Communications Societies (INFOCOM 2002)*, Vol. 3, pp. 1567–1576.

Ye W, Heidemann J, Estrin D (2004) Medium access control with coordinated adaptive sleeping for wireless sensor networks. *IEEE/ACM Trans Netw* 12(3):493–506.

Yu Y, Govindan R, Estrin D (2001) Geographical and energy aware routing: A recursive data dissemination protocol for wireless sensor networks. UCLA, Technical Report, UCLA/CSD-TR-01-0023.

Chapter 6
Standards and Safety Regulations for WSNs

Abstract The use of any technology requires that its users be confident of both its usefulness and its safety. In an effort to guarantee this, standardization bodies around the world generate standards to be followed by product manufacturers. When dealing with electronic communications systems such as WSNs, issues such as electromagnetic compatibility, the safety of their operation, the confidentiality and security of private information, and environmental awareness are of great practical importance. This chapter reviews how these safety issues apply to WSNs and presents some of the significant European regulations on this matter together with some notes on the open issues that may have to be treated in possible future standards.

6.1 Introduction to the Regulatory Aspects of WSNs

In this chapter, we consider the main European regulation applicable to WSNs. This regulation includes aspects such as electromagnetic compatibility and health risks related to the production, use, and disposal of corresponding equipment. It has to be noted, however, that this regulation applies only to Europe; other regions have their own legislation on such matters. European laws provide good guidance on the most significant aspects in terms of the safety and regulation of WSN systems.

The R&TTE directive (Dir 1999/5/EC) of the European Union governs the R&TTE (radio equipment and telecommunications terminal equipment) market in Europe to allow for free movement of these products while assuring that under normal use there is an efficient use of spectrum, no harmful interferences, and no avoidable health risks. It also mentions the necessity of assuring the privacy of personal data. However, it is not within the scope of this directive to harmonize the radio-frequency spectrum. In fact, the spectrum is not fully harmonized throughout Europe, as the Member States have individual authority on the spectrum usage matter.

WSN nodes fall within the scope of the R&TTE directive since they may be considered "radio equipment," whose definition is "a product, or relevant

A.-B. García-Hernando et al., *Problem Solving for Wireless Sensor Networks*,   153
DOI: 10.1007/978-1-84800-203-6_6, © Springer-Verlag London Limited 2008

component thereof, capable of communication by means of the emission and/or reception of radio waves utilising the spectrum allocated to terrestrial/space radiocommunication." The frequency bands used by WSN evidently are considered inside the limits of radio waves.

The essential requirements that the directive states all apparatus must follow are (including textual quotes from the directive and the most recent legislation references at the time of this writing)

- "The protection of the health and the safety of the user and any other person," including the safety objectives of the so-called low-voltage directive (LVD) (Dir 2006/95/EC), although without the voltage limit
- "The protection requirements with respect to electromagnetic compatibility contained in" the so-called electromagnetic compatibility (EMC) directive (Dir 2004/108/EC)
- The efficient use of "the spectrum allocated to terrestrial/space radio communication and orbital resources so as to avoid harmful interference"

Apparatus that comply with the relevant essential requirements of this directive will show the "CE" conformity marking. There are also directives related to environmental protection when manufacturing, using, or disposing of electronic devices; we will present these later in this chapter.

European directives do not usually contain technical expressions for their requirements. Instead, this expression is usually found in European standards (EN), which carry the obligation of being implemented at the national level in all member countries. Three European standards organizations may produce European standards: CENELEC (European Committee for Electrotechnical Standardization), CEN (European Committee for Standardization), and ETSI (European Telecommunications Standards Institute). Products that comply with the harmonization standards under a certain directive are presumed also to comply with the associated directive requirement(s). Thus, European standards together with the directives they refer to give complete meaning to the regulatory issues applicable to products on the European market.

WSNs, although within the scope of R&TTE and other directives, constitute a relatively new technology, especially regarding commercial products and applications. Thus, some issues are being identified that may in the future require specific regulation for WSNs and similar systems (i.e., ubiquitous wireless systems). This chapter also reviews currently open issues that could cause the revision of safety and regulatory provisions for these types of systems.

Table 6.1 summarizes the main safety and regulatory issues that have to be dealt with in WSN systems. It also gives an overview of the structure of the rest of the chapter.

Table 6.1 Safety and Regulatory Issues Related to Several Aspects of WSN

Aspect of WSN	Safety and Regulatory Issues
Electromagnetic radiation	Electromagnetic compatibility with other devices in their surroundings
	Biological effects of exposure to radiation
Materials used in nodes and batteries	Environmental impact and health risks
Data processed, sent, and stored	Data security and privacy

6.2 Electromagnetic Compatibility

Depending on the application, WSN nodes may have to share their surroundings with other electronic devices (e.g., computers, domestic appliances, or medical equipment). Both WSN nodes and the other devices may generate electromagnetic radiation and/or be affected by it. It is clear that for WSNs to be of any utility, they must be able to function properly in the electromagnetic environment in which they are supposed to be placed. This includes, under normal use, not interfering with the functioning of other devices and having a sufficient level of immunity to the radiation emanating from that other equipment.

The protection requirements present in the electromagnetic compatibility (EMC) directive (Dir 2004/108/EC) apply to R&TTE, and thus to WSNs. These requirements can be summarized very briefly as the "no interference" and "immunity" properties of the equipment with respect to electromagnetic radiation. Quoting from the directive,

> Equipment shall be so designed and manufactured, having regard to the state of the art, as to ensure that:
> (a) the electromagnetic disturbance generated does not exceed the level above which radio and telecommunications equipment or other equipment cannot operate as intended;
> (b) it has a level of immunity to the electromagnetic disturbance to be expected in its intended use which allows it to operate without unacceptable degradation of its intended use.

The nature of the involved physical phenomena and the high number of electromagnetic environments that a product may have to support make it difficult to design tests that prove the compliance with the two requirements stated above. To help in this matter, there are European EMC standards under the EMC directive that specify tests and limits to prove that the requirements are met. It is not compulsory to use these standards to prove compliance with the directive, although it is the recommended procedure whenever appropriate standards exist.

There are three types of EMC standards (CENELEC, 2005). Basic standards specify mainly tests and measurements to be done, but they do not contain prescribed limits. To find concrete EMC requirements (including limits), generic or product standards have to be used, which refer to basic standards when necessary. If a product standard exists for the product or product family, it has preference over the applicable generic standard.

In general, more than one EMC standard has to be verified to prove that a product is compliant with the EMC directive. To begin with, both the "emission" and the "immunity" parts of the requirements have to be assessed. Also, if the product has several functionalities or functioning modes, or if it is to be used in different environments, more than one standard may have to be included to cover all the possibilities of normal use.

All this makes the process of deciding what standards to apply to a product not always straightforward. It is even more difficult if the product to be evaluated is new.

In the case of WSN radio devices, depending on their frequency band, range, power level, and modulation technique, the following EMC standards may be applicable (generated by the ETSI):

- EN 300 220-1 to 3: electromagnetic compatibility and radio spectrum matters (ERM); short-range devices (SRD); radio equipment to be used in the 25-MHz to 1000-MHz frequency range with power levels ranging up to 500 mW.
- EN 300 440-2: electromagnetic compatibility and radio spectrum matters (ERM); short-range devices; radio equipment to be used in the 1-GHz to 40-GHz frequency range. Part 2: Harmonised EN under Article 3(2) of the R&TTE directive.
- EN 300 328: electromagnetic compatibility and radio spectrum matters (ERM); wideband transmission systems; data transmission equipment operating in the 2.4-GHz ISM band and using wideband modulation techniques.
- EN 301 489-03: electromagnetic compatibility and radio spectrum matters (ERM); electromagnetic compatibility (EMC) standard for radio equipment and services; Part 3: specific conditions for short-range devices (SRD) operating on frequencies between 9 kHz and 40 GHz.

For instance, Standard EN 300 220 (SRD operating at frequencies up to 1 GHz) establishes both transmitter and receiver parameters. Depending on how critical the application is, the receiver will have to comply with a different number of limits. The standard covers all types of modulation, both narrowband and wideband (including spread-spectrum technologies in the latter), with some limits applicable to only a subset of them. Examples of transmitter requirements include maximum "effective radiated power" and limitations on "transient power," defined as "the power falling into adjacent spectrum due to switching the transmitter on and off during normal operation." This latter limit may be especially significant in WSNs since, in order to lower consumption, the transmitter is not continuously on. Among the receiver requirements we can find the "maximum usable sensitivity" (the minimum level of signal at the receiver in order to obtain a certain quality) and the "adjacent channel selectivity" (which measures the capability of the receiver to operate correctly in the presence of an unwanted signal located in an adjacent channel).

Achieving electromagnetic compatibility for tiny devices as WSN nodes may be complex, since the miniaturization of the devices makes it difficult to avoid interference even between the electronic components of the node itself. This could prove to be a challenge to keep decreasing the nodes' size.

6.3 Biological Effects of Radiation

The exposure to EMF (electromagnetic fields), including, of course, RF (radio-frequency) radiation, is known to cause potentially harmful biological effects on humans. These biological effects are classified into thermal and nonthermal effects:

- Thermal effects are agreed to cause unwanted biological effects. They appear when the heat induced by the radiation exposure is higher than what the natural body circulation may drain.
- Nonthermal effects have a much lower level of agreement and understanding. They comprise all the biological effects caused by radiation exposure that are independent of the temperature rise. Although there is no totally agreed scientific evidence on the harmfulness of these effects, there is no evidence on their innocuousness either. Thus, precautionary measurements have to be taken when advisable given the potential risk.

Issues regarding the biological effects of exposure to EMF are covered under the low-voltage directive (LVD) (Dir 2006/95/EC). Inside its framework, the Council of the European Union has published, on the basis of the vast amount of scientific documentation on this matter, and with the advice of the International Commission on Non-Ionising Radiation Protection (ICNIRP), Recommendation 1999/519/EC "on the limitation of exposure of the general public to electromagnetic fields (0 Hz to 300 GHz)" (Rec 1999/519/EC). It contains basic limits to prevent harmful exposure to EMF, derived only from threshold values that have been proven to cause acute harmful effects. Logically, there is a safety factor to adjust these thresholds to the restrictions present in the recommendation, in this way giving some protection against the possible long-term effects.

The basic restrictions present in the recommendation are expressed in terms of the magnetic flux density, current density, specific energy absorption rate (SAR), and/or power density depending on the frequency of the field. The frequency bands used by current WSN platforms range between 300 MHz and 2.4 GHz (i.e., they use frequencies inside the UHF band). Thus, the applicable frequency range of Recommendation 1999/519/EC is 10 MHz–10 GHz, for which the basic restrictions are expressed in terms of the SAR. There are three SAR figures to consider:

- The limit on the whole body average SAR is 0.08 W/kg.
- The limit on the localized SAR for head and trunk is 2 W/kg.
- The limit on the localized SAR for limbs is 4 W/kg.

SAR is defined in the recommendation as "the rate at which energy is absorbed per unit mass of body tissue and is expressed in watts per kilogram (W/kg)." The whole body average SAR is useful for evaluating the adverse thermal effects of exposure, while the localized SAR figures are necessary for cases in which small parts of the body may be especially exposed to the radiation.

The power emitted by WSN may be considered below the established safety limits. However, if nodes are placed on the human body and they are meant to continuously monitor and send data, the need to investigate the possible biological effects arises again, since the local SAR may be high. Note, for instance, the following warning present in the datasheet of a commercial mote product: "If the module will be used for portable applications, the device must undergo SAR testing" (Moteiv, 2006). And, in fact, the sensing of biological or medical parameters using WSN nodes located on (or even inside) the body is an application probably not uncommon in the future. See Mailhes et al. (2002) and Reeves et al. (2006) for two of the many examples that can be found on this kind of application.

The research on the possible biological effects of WSN radiation is very recent, as most existing research on WSNs is targeted to other technical issues such as energy efficiency or quality of service. However, it is possible to find some papers that investigate how to design protocols and algorithms for WSNs that take into account the effects of radiation exposure on humans. For the reasons stated above, such research focuses on biological wireless sensor networks in which the sensor nodes are placed directly on the human body. See, for instance, Tang et al. (2005), who propose an algorithm for selecting the cluster heads of an implanted biosensor network in a way that minimizes the increase in temperature in human tissue.

More recently, Ren and Meng (2006) proposed an "equivalent Coefficient-of-Absorption-and-Bioeffects" to evaluate the possible biological effects of a WSN that measures biological data. The novelty in this case comes from the consideration of not only the thermal effects (as in Tang et al. (2005)) but also the nonthermal effects. The value of this coefficient depends on the physical output power of the transmitters, the incident power density, the network traffic load, and the tissue characteristics (certain parts of the body are known to have a higher sensitivity to RF exposure than others). Following this research, Ren and Meng (2006) indicate some design parameters for the network protocols of a WSN aimed at reducing their biological effects, which include power control and rate control, and also propose a rate control algorithm to minimize these effects.

Clearly, analyzing the biological effects of the radiation from WSNs is a field in which a lot of future research can (and probably should) be done. If current legislation proves to be insufficient for the application of this new technology, scientific studies will be necessary as a basis of future regulation.

6.4 Environmental Impact

WSNs, like any other electrical or electronic product, may cause negative environmental effects that can be classified into the following categories (Köhler and Erdmann, 2004):

- Global resource depletion: The production of semiconductor-based electronic products consumes large quantities of natural resources.
- Energy use: The electricity consumption of electronic devices is a significant portion of the total figure. In WSNs, those nodes connected to mains and those with rechargeable batteries consume electricity.
- Production of dangerous substances: Several components of devices and batteries may have health hazards and/or produce pollution if they are not properly limited and treated or recycled. These substances may be present during the production, use, and disposal of the products. Moreover, failing to recycle these products leads to the worsening of the first and second effects since obtaining and preparing new raw materials is less efficient than recycling them. This is especially (although not exclusively) true for batteries, currently present in every WSN node not connected to mains.

To cope with these risks, the European Union has produced regulations whose goal is to minimize the environmental hazards of electric and electronic equipment sold in Europe. In principle, nothing prevents WSNs from falling into the application scope of these directives, summarized as follows:

- The **WEEE** directive (Dir 2002/96/EC) aims to reduce the negative environmental effects caused by waste electrical and electronic equipment by (1) reducing equipment disposal (to do so, WEEE promotes the recycling and reuse of the devices) and (2) reducing the environmental damages caused by the processes carried out during their life cycle (improving their environmental performance).
- The **RoHS** directive (Dir 2002/95/EC) restricts the use of substances in electrical and electronic equipment that are known to be hazardous for human health and the environment. RoHS does not apply to batteries. This directive mandates that "Member States shall ensure that, from 1 July 2006, new electrical and electronic equipment put on the market does not contain lead, mercury, cadmium, hexavalent chromium, polybrominated biphenyls (PBB) or polybrominated diphenyl ethers (PBDE)" (Dir 2002/95/EC). Later, several Commission Decisions adopted during 2005 and 2006 amended the RoHS directive to establish applications exempted of the aforementioned requirement and to indicate the maximum concentration

of the hazardous substances tolerable in homogeneous (of uniform composition) materials.

- There is a European directive entirely devoted to **batteries** (Dir 2006/66/EC), since all batteries contain hazardous materials (e.g., mercury, lead, or cadmium) and metals that may be recycled (e.g., nickel, cobalt, or silver) with the consequent reduction in the needed energy with respect to having to extract and prepare virgin metals. Thus, this directive's goal is to promote the recycling of batteries and accumulators as well as restrict certain dangerous substances in their composition. This new directive, published on September 26, 2006, replaces the current directive on batteries and includes a wider range of products.

Although this is European regulation, a manufacturer that wants to sell products both inside and outside the EU will probably choose not to have different manufacturing processes for the different regions, thus applying the EU directives to its entire production (Eveloy et al., 2005).

These directives consider exceptions (products to which they do not apply) that are more related to the application of the product (e.g., military, emergency, or medical uses) than to the characteristics of the product itself. This could be taken into account for the WSNs used for these particular types of applications. Apart from the aforementioned exceptions, it seems that the placing of unattended WSN nodes in an area, e.g., to measure environmental parameters, has to be done in a way that ensures that the possibility will exist for their proper treatment at the end of their lifetime. However, if a truly ubiquitous presence of WSN-like systems becomes a reality, it will be very difficult to ensure that there is no loss of valuable materials.

Having said this, it should be noted that WSNs may also produce beneficial environmental effects. In fact, many of their potential applications are precisely designed to indirectly obtain some sort of environmental benefits. These are what may be called second-order environmental impacts, as opposed to the first-order impacts hitherto described (Köhler and Erdmann, 2004). A few of these positive effects from the use of WSNs may be exemplified as follows:

- The use of WSNs to help manage and control industrial processes may improve their efficiency and make them consume fewer resources and energy.
- The same is true for WSNs used to control homes or buildings (e.g., WSNs that help manage the heating system in terms of efficiency).
- WSNs used to give prompt information on the state of roads could allow motorists to choose faster and more efficient itineraries, thus reducing pollution and fossil fuel use.
- Many WSN nodes have to operate with no connection to the mains and without human intervention. These circumstances are incentive for the research on alternative energy sources, such as solar panels or energy-harvesting methods, that prolong the life of the nodes with respect to the use of batteries. Research literature already exists on this. See, for instance,

the work by Jiang et al. (2005) in which a system composed of a solar panel, a super capacitor, and a rechargeable battery is used to feed a Telos mote. With the correct charging control strategy, the authors claim to largely prolong the life of sensor networks (or even make it perpetual under certain operational boundaries). Even commercial energy-harvesting modules are beginning to appear that are suitable for WSNs and the like.

- The same reasons mentioned in the previous bullet (the autonomous operation of the nodes) also promote research on more efficient energy use by WSN nodes.
- Even well-known clean energy sources for everyday life may benefit from WSNs, since these provide better means to manage them, and the control technologies needed to make an optimal use of them are complex.

Second-order beneficial effects could compensate for the first-order harmful effects of WSNs, but only if the usage of the technology is in the framework of adequate economic structures and lifestyles. Otherwise, so-called third-order adverse effects may appear that effectively invalidate the second-order benefits, something that has happened with most of the preceding technologies (Köhler and Erdmann, 2004). In order to prevent this from happening, new regulations could be needed from the first stages of the adoption of WSN and similar pervasive technologies.

6.5 Data Security and Privacy

Some of the applications of WSNs involve the exchange of potentially private information regarding people or business processes. The ubiquity of these devices and their ability to collect and wirelessly transmit a great variety of data make the security of their operation an important issue. This is especially true when dealing with medical applications (e.g., sending patients' medical signals obtained by wearable sensors) or in general when involving people (e.g., tracking the position or detecting the crossing of an invisible perimeter by humans).

In fact, the security and privacy of personal data are currently considered fundamental rights in most developed countries. This way, there exist national data protection bodies in charge of enforcing this right inside their respective region or country.

Obviously, the introduction of a new technology such as WSNs must not weaken the safety and rights of people, not only physically but also regarding their right to maintain the privacy and security of their personal data. Only if people perceive that their rights are being guaranteed by proper regulations (together with the corresponding means to enforce it) will the new applications be generally accepted.

The security and privacy of data in applications that electronically process, store, or transmit information are not new issues, and corresponding regulations already exist. See, for instance, the EC 2002 directive in a European context.

The novelty introduced by WSNs and other technologies (e.g., RFID—Radio Frequency Identification) is curiously also one of their strengths: the ubiquity and nonintrusiveness of their operation, which makes the process of recollecting and sending data practically unnoticeable to their users.

As a sign of the importance and complexity of this matter, the European Data Protection Supervisor (EDPS) has classified as being of particular interest three subjects related to RFID in which the privacy of data may be compromised (IssuesEDPS). Although related to RFID and not explicitly to WSNs, these subjects raise issues that are common to both technologies. For instance, EDPS states that "Special emphasis is laid on the potential privacy, security and integrity issues implicated by the use of RFID technology to identify and track individuals" (EDPS 2007). Since the tracking of individuals is also an important application of WSNs, many of these issues will also apply to them. Moreover, some researchers already see RFID and WSN as converging technologies (see, for instance, Harrop (2006)). And, quoting again from the EDPS, it should not be forgotten that "RFID together with biometrics, ambient intelligence environments and Identity Management Systems" have been identified by EDPS "as technological developments that are expected to have a major impact on data protection" (Hustinx, 2007).

Whether the existing regulation on data protection is enough for WSNs or a new legal framework for WSN (possibly together with similar technologies) is necessary is a complex matter. One of the functions of EDPS is precisely to advise stakeholders of the applicability of the existing regulation to new technologies and also to detect the need for new legal support. It is outside the scope of this section to perform a detailed analysis on this issue, although it is the opinion of the authors that the importance of taking this into account had to be stressed.

References

CENELEC Home page. http://www.cenelec.org/Cenelec/Homepage.htm. Accessed December 2007.

CENELEC (2005) CENELEC guide no. 24: Electromagnetic Compatibility (EMC) Standardization for Product Committees.

The European Parliament and the Council of the European Union (1999) Directive 1999/5/EC of the European Parliament and of the Council of 9 March 1999 on radio equipment and telecommunications terminal equipment and the mutual recognition of their conformity. *Off J Eur Communities*, L, Legis 91 of 7. 4. 1999:10–28.

The European Parliament and the Council of the European Union (2003) Directive 2002/95/EC of the European Parliament and of the Council of 27 January 2003 on the restriction of the use of certain hazardous substances in electrical and electronic equipment. *Off J Eur Communities*, L, Legis L 37 of 13.2.2003:19–23.

The European Parliament and the Council of the European Union (2003) Directive 2002/96/EC of the European Parliament and of the Council of 27 January 2003 on waste electrical and electronic equipment (WEEE). *Off J Eur Communities*, L, Legis 37 of 13.2.2003:24–38.

The European Parliament and the Council of the European Union (2004) Directive 2004/108/EC of the European Parliament and of the Council of 15 December 2004 on the approximation of the laws of the Member States relating to electromagnetic compatibility. *Off J Eur Communities*, L, Legis 390 of 31.12.2004:24–37.

The European Parliament and the Council of the European Union (2006) Directive 2006/66/EC of the European Parliament and of the Council of 6 September 2006 on batteries and accumulators and waste batteries and accumulators. *Off J Eur Communities*, L, Legis 266 of 26.9.2006:1–14.

The European Parliament and the Council of the European Union (2006) Directive 2006/95/EC of the European Parliament and of the Council of 12 December 2006 on the harmonisation of the laws of Member States relating to electrical equipment designed for use within certain voltage limits. *Off J Eur Communities*, L, Legis 374 of 27.12.2006:10–9.

Directive 2002/58/EC of the European Parliament and of the Council of 12 July 2002 concerning the processing of personal data and the protection of privacy in the electronic communications sector (directive on privacy and electronic communications), amended by Directive 2006/24/EC of the European Parliament and of the Council of 15 March 2006.

Home page of the European Data Protection Supervisor. http://www.edps.europa.eu/EDPS-WEB/edps/lang/en/. Accessed December 2007.

Eveloy V, Ganesan S, Fukuda Y, et al. (2005) Are you ready for lead-free electronics? IEEE Trans Compon Packag Technol 28(4):884–94.

Harrop P (2006) Active RFID 2006–2016. IDTechEx.

Hustinx P (2007) Opinion of the European Data Protection Supervisor on the communication from the Commission to the European Parliament, the Council, the European Economic and Social Committee and the Committee of the Regions on Radio Frequency Identification (RFID) in Europe: Steps towards a policy framework COM(2007) 96. EDPS.

EDPS (2007) Inventory of issues of particular interest to EDPS, Last update: December 2007. http://www.edps.europa.eu/EDPSWEB/webdav/site/mySite/shared/Documents/Consultation/Priorities/07-12-20_Inventory_EN.pdf. Accessed December 2007.

Jiang X, Polastre J, Culler D (2005) Perpetual environmentally powered sensor networks. In *Proceedings of the 4th International Symposium on Information Processing in Sensor Networks (IPSN 2005)*, pp. 463–468.

Köhler A, Erdmann L (2004) Expected Environmental Impacts of Pervasive Computing. *Hum Ecol Risk Assess* 10(5):831–52.

Mailhes C, Castanié F, Henrion S, et al. (2002) The U-R-Safe project: An innovative multidisciplinary approach for an "anywhere" care of the elderly. In *Proceedings of the 2nd International Telemedicine Symposium*.

Moteiv Corporation (2006) Tmote Sky: Datasheet (11/13/2006).

The Council of the European Union (1999) Council recommendation (1999/519/EC) of 12 July 1999 on the limitation of exposure of the general public to electromagnetic fields (0 Hz to 300 GHz). *Off J Eur Communities*, L, Legis 199 of 30. 7. 1999:59–70.

Reeves AA, Ng JWP, Brown SJ, et al. (2006) Remotely supporting care provision for older adults. In *Proceedings of the International Workshop on Wearable and Implantable Body Sensor Networks (BSN 2006)*.

Ren H, Meng MQH (2006) Rate control to reduce bioeffects in wireless biomedical sensor networks. In *Proceedings of the 3rd Annual International Conference on Mobile and Ubiquitous Systems: Networking & Services*, pp. 1–7.

Tang Q, Tummala N, Gupta SKS, et al. (2005) Communication scheduling to minimize thermal effects of implanted biosensor networks in homogeneous tissue. *IEEE Trans Biomed Eng* 52(7):1285–94.

Chapter 7
European Research Projects Related to WSNs

Abstract We summarize some relevant research European projects to give the reader a picture on the work in progress in the industrial and academic communities regarding WSNs. For each project we have included a list with basic information (useful if more data on the project are sought) and a summary of the main project objectives. The large number of projects whose research objectives fall into WSNs (in whole or in part) also gives an idea of the high current interest in this technology.

7.1 UbiSec&Sens

- Name: Ubiquitous Sensing and Security in the European Homeland
- Program: FP6-IST (STREP)
- URL: http://www.ist-ubisecsens.org

Summary of the Project's Objectives

The main goals of the project are to develop an architecture and new design cycle for secure sensor networks and to develop a toolbox of components designed with security in mind. The targets of these developments are medium and large wireless sensor networks. The project tests some scenarios in order to check these networks' needs regarding issues such as scalability, security, reliability, self-healing, and robustness. Also, the intersection of the security issues with routing and in-network processing is a research issue included in the project.

One of the interesting outcomes of this project is the document entitled "Scenario definition and initial threat analysis" (Casaca and Westhoff, 2006) that explains the security threats in WSNs.

7.2 CoBIs

- Name: Collaborative Business Items
- Program: FP6 STREP Project # IST 004270
- URL: http://www.cobis-online.de/index.html

A.-B. García-Hernando et al., *Problem Solving for Wireless Sensor Networks*, DOI: 10.1007/978-1-84800-203-6_7, © Springer-Verlag London Limited 2008

Summary of the Project's Objectives

To quote this project's Website, its main goal is "to apply networked embedded systems technologies in large-scale business processes and enterprise systems by developing a platform for directly handling processes at the relevant point of action rather than in a centralized back-end system."

This project aims to provide new capabilities to usual objects in terms of monitoring or communication using a WSN infrastructure. In fact, sensor and communication possibilities are integrated into these objects. By monitoring object properties, either their own internal state or their direct environment may be observed. Communication properties permit different collaborations (e.g., object to object or object to user) for new organizational or functional purposes.

A demonstration of this system was presented at the IST 2006 conference (IST, 2006). It applied to the management of hazardous chemicals containers. The demonstration scenario is the following: If two containers that contain two incompatible substances are placed too close to one another, they can trigger an alarm and prevent any possible accident. Another scenario consists of preventing unqualified people from using potentially hazardous equipment. This would be achieved by putting sensors on workers' suits to check whether a person meets certain conditions to access a certain area of a factory. The nodes would then communicate with other nodes in the vicinity, either on clothes or on equipment, to verify the worker's access rights. The access is authorized only if all prerequisites are met.

The considered applications include adding intelligence (monitoring and communication) to items that already could be traced by means of RFID technologies. These objects are capable of communicating in a peer-to-peer manner or with Enterprise Resource Planning (ERP) software.

7.3 WINNER

- Name: Wireless World Initiative New Radio
- Program: FP6 integrated project
- URL: https://www.ist-winner.org/index.html

Summary of the Project's Objectives

The WINNER project proposes to develop a future radio access system with improved performance and interoperability (i.e., to enable the connection and cooperation with various wireless systems in optimal conditions). The radio interface is supposed to be generic and flexible enough to adapt itself to any kind of user or application requirement (e.g., various kinds of network topologies, hardware technologies, and frequency-sharing systems). The aimed peak data rate is up to approximately 100 Mbps for the new mobile access and 1 Gbps for new nomadic/local area wireless access. Work focuses on technologies, topologies, scalability, wide operational and application range, radio channel model development, and optimal use of the frequency spectrum.

The first phase of the WINNER project ended on December 31, 2005, after two years of work. The work focused on the definition and study of new technologies for the future radio interface. The second phase began on January 1, 2006, also for two years, to optimize the concept and provide some real-life proof to demonstrate the performances (public deliverables are accessible on the Website).

WINNER is part of the Wireless World Initiative (WWI Home page), "a major European research initiative to create the technologies needed for Systems Beyond 3 G (B3G) in a cross industry and academia research collaboration." It was established in 2002 to prepare the submission of FP6 European integrated projects to begin in January 2004. The objectives of these projects were oriented toward system architecture, users' requirements, quality of service, security, resilience, reconfigurability, operability, and validation of systems.

7.4 AWARE

- Name: Platform for Autonomous Self-Deploying and Operation of Wireless Sensor-Actuator Networks Cooperating with Aerial Objects
- Program: FP6 STREP Project # IST-2006-33579
- URL: http://grvc.us.es/aware

Summary of the Project's Objectives

AWARE provides software tools for middleware embedded on flying objects (unmanned aerial vehicles, or UAV) or humans, with the ability to carry sensors and establish cooperation and communication between each other and with earth platforms. The network created should be self-deployed, autonomous, and able to work in places with low accessibility or places without a dedicated communication infrastructure. Figure 7.1 shows an application scenario example considered by the AWARE consortium.

The considered application domains are civil security and disaster management and filming dynamically evolving scenes with mobile objects; three experiments were carried out during the project.

The sensors carried by the UAVs can be cameras or other light and low-power sensors. GPS will also be used to attain autolocalization.

7.5 Sensation

- Name: Advanced Sensor Development for Attention, Stress, Vigilance and Sleep/Wakefulness Monitoring
- Program: FP6 integrated project
- URL: http://www.sensation-eu.org/

Fig. 7.1 Example of an
AWARE application
scenario. (Courtesy of the
AWARE consortium;
http://www.aware-project.
net.)

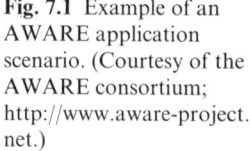

Summary of the Project's Objectives

The SENSATION project's main objective is to study and provide technological solutions, in the form of wireless sensor arrays, to measure physiological data. Physiological measurement is used in order to detect and predict unwanted sleep phases. The systems must be low-cost and noninvasive and allow the real-time detection of the possible symptoms of an upcoming sleep phase.

Specific sensors are developed during the project, and some applied studies on brain operation are achieved.

The application domains range from hypovigilance in transportation systems (automotive, aviation) to security improvement in potentially dangerous industries (prevention of serious industrial accidents due to excessive fatigue).

Subproject 2, called "Micro and nano sensor development," is dedicated to the development of novel sensors and their wireless network interconnection. It is the part of the project most related to WSNs. To quote from the Website, this subproject is devoted to

> developing a wide array of novel micro and nano sensors for unobtrusive monitoring of human physiological state and activity; their interconnection through an embedded connectivity at the body, local and wider area; and their integration in multi-sensorial systems through innovative signal processing and computational intelligence algorithms for data fusion, data management and power consumption minimization.

The project does not address low consumption.

7.6 e-SENSE

- Name: Capturing Ambient Intelligence for Mobile Communications Through Wireless Sensor Networks
- Program: FP6 IST integrated project IST-FP6-IP-027227
- URL: http://www.ist-e-sense.org

Summary of the Project's Objectives

The e-SENSE project's objective is to develop a broad and generic framework to enable the convergence of heterogeneous wireless communication systems for an ambient intelligence context. This framework is supposed to be accessible to any kind of wireless network in terms of size, scale, composition, mobility, and application domain. Sensor networks can be local (body sensor network), large-range (object-to-object interaction), or wide area (environment monitoring), or any combination of these, aimed at integration into future "Beyond 3 G" (B3G) systems.

According to the project's Website, the project's mission is to provide

heterogeneous wireless sensor network solutions to enable context capture for ambient intelligence, in particular for mobile and wireless systems beyond 3 G, thus enabling truly multi-sensory and personal mobile applications and services, as well as assisting mobile communications through sensor information.

The main research domains of the project are as follows:

- Efficient wireless communications (WP3)
- Scalable and reconfigurable transport of data (WP2&4)
- Distributed processing middleware (WP4)

7.7 WASP

- Name: Wirelessly Accessible Sensor Populations
- Program: FP6-IST
- URL: http://www.wasp-project.org/

Summary of the Project's Objectives

The main goal of the project is the creation of a complete system view of the construction of large populations of collaborative objects. With this goal in mind, the project intends to encourage industry to use the results of academic research, by covering all levels of the wireless sensor network, from the application to the node and the network. The validation of the results is done in three different business areas: road transport, elderly care, and herd control. All of these areas were selected because of their social significance as well as the complex nature of the projects, specifically the large range of requirements.

7.8 MIMOSA

- Name: Microsystems Platform for Mobile Services and Applications
- Program: FP6-IST
- URL: http://www.mimosa-fp6.com/

Summary of the Project's Objectives

The project aims to develop a mobile phone–centric, open-technology platform to make ambient intelligence a reality. The project is expected to identify and develop generic microsystem blocks for ambient intelligence. Microsystems are used due to their low cost, low power consumption, and small size. The identified and developed blocks provide the project with the required functionality for the local connectivity, context sensing, intuitive user interface, energetic autonomy, and microsystem integration technology.

7.9 E2R

- Name: End-to-End Reconfigurability
- Program: FP6-IST
- URL: http://e2r2.motlabs.com

Summary of the Project's Objectives

The main goal of the E2R project is to invent, develop, test, and show the results of an architectural design of reconfigurable devices and supporting system functions, with the aim of offering a wide set of operational options to users, applications, service regulators, operators, and providers in the context of heterogeneous systems. An end-to-end perspective should drive developments and research in order to achieve the above-mentioned goal. End-to-end reconfigurable systems provide and execute environments that, through the use of cognitive methods, are able to optimize the usage of resources and obtain versatility. With all these features, the final user should receive the required service when and where needed and at an affordable price.

7.10 CRUISE

- Name: Creating Ubiquitous Intelligent Sensing Environments
- Program: FP6-IST Network of Excellence
- URL: http://www.ist-cruise.eu/

Summary of the Project's Objectives

This project's intention is to be the focal point in the planning and coordination of research in the communication and application aspects of WSNs in Europe. A crucial part of the project is the creation of a state-of-the-art knowledge base available to the general public, by collecting, comparing, validating, and disseminating information. The research is focused on the solution of specific theoretical and technical problems that may allow the construction of sensor network applications that have the potential to significantly affect European society. The project also hopes to stimulate discussion on standardization, international collaboration, and intellectual property. Another main objective involves teaching and training results regarding wireless sensor networks as well as seeking new techniques to teach and train these technologies.

7.11 RUNES

- Name: Reconfigurable Ubiquitous Networked Embedded Systems
- Program: FP6-IST integrated project
- URL: http://www.ist-runes.org

Summary of the Project's Objectives

The main objective is to enable the creation of a large-scale and widely distributed heterogeneous network of embedded systems capable of adapting and interoperating in their environments. In order to achieve widespread usage of network embedded systems, this project develops a standardized architecture capable of self-organization and adaptation to a changing environment. Also among the outcomes of the project, an adaptive middleware platform and a series of application development tools are created to allow the developers to flexibly interact with the environment and at the same time offer a level of abstraction that will facilitate the development and usage of applications. This reduces the development costs as well as the time to market of the application. The project also examines potential uses of the technology, developing demonstrator systems and designing training courses to show the benefits of the obtained results.

7.12 Smart Messages

- Name: Smart Messages Project
- Program: NA
- URL: http://discolab.rutgers.edu/sm/

Summary of the Project's Objectives

The main goal of the Smart Messages Project (Smart Messages) is to develop a network architecture for large-scale embedded systems (NES) whose functioning is not supervised by humans. NES are characterized by restriction of resources, heterogeneity, and volatile nodes. These unique characteristics make traditional distributed computing models not easy to use in such networks. In order to solve this problem, the Smart Messages Project proposes a distributed computing model called Cooperative Computing, which is based on execution migration. In this model, applications are composed of dynamic collections of Smart Messages (SMs), and each node cooperates by providing a common system support.

Two papers that have been published in the context of this project are the following: "Cooperative computing in sensor networks" (Iftode et al., 2004) and "Smart Messages: A distributed computing platform for networks of embedded systems" (Kang et al., 2004). The main concept treated in both documents is the cooperative computing model. In this model, a distributed application can be developed without a priori knowledge of the scale and topology of the network, or the specific functionality of each node. To prove that any protocol or application can be written using SMs, two previously proposed applications have been implemented: SPIN (Heinzelman et al., 1999; Kulik et al., 2002) and directed diffusion (Intanagonwiwat et al., 2000).

The project has also included a node prototype (see Fig. 7.2) and two application prototypes: "EZCab: An automatic system for booking a cab in cities" (Zhou et al., 2004) and "TrafficView: A scalable traffic monitoring system," (Disco Lab) which were evaluated with simulations.

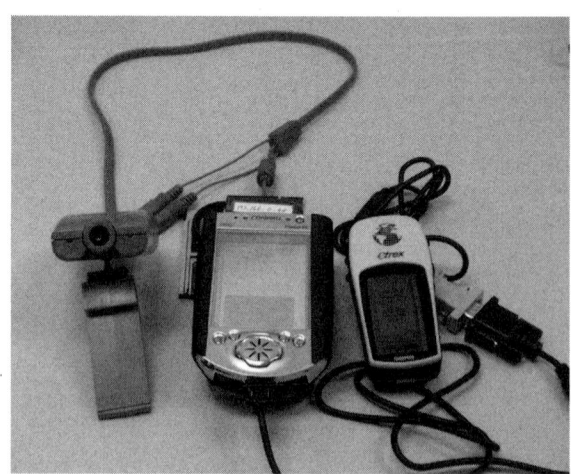

Fig. 7.2 "Smart Messages" prototype node with camera and GPS attached. (From [Iftode et al., 2004].)

7.13 EYES

- Name: Energy Efficient Sensor Networks
- Program: IST-2001-34734
- URL: http://www.eyes.eu.org/

Summary of the Project's Objectives

The main goal of the EYES project is to develop a new smart node technology needed for building self-organizing and collaborative sensor networks. As the project's description reads, "This technology will enable the creation of a new generation of sensor nodes, which can effectively network together so as to provide a flexible platform for the support of a large variety of mobile sensor network applications" (Havinga et al., 2003). The feasibility of the concepts and technologies theoretically developed is demonstrated with a sensor network prototype in which some example applications are deployed.

The EYES architecture is defined over two distinct key system layers of abstraction: the sensor and networking layer on one side, and the distributed services layer on the other. The sensor and networking layer contains the sensor nodes (the physical sensor and wireless transmission modules) and the network protocols. Communication protocols developed in this layer take into account the mobility of nodes and dynamic changes of topology. On the other hand, the distributed services layer contains distributed services to support mobile sensor applications. There are two major services: the lookup service, which controls the node self-configuration, and the information service, which deals with the collection of data.

7.14 Embedded WiSeNts

- Name: Cooperating Embedded Systems for Exploration and Control Featuring Wireless Sensor Networks
- Program: FP6-IST-004400
- URL: http://www.embedded-wisents.org

Summary of the Project's Objectives

The main goal of the Embedded WiSeNts project is to connect the classic concept of embedded systems with two more recent system concepts: ubiquitous computing and wireless sensor networks. Embedded WiSeNts proposes that "these three types, of actually quite diverse, state-of-the-art systems share some principal commonalities but also share some complementary aspects that make a combination of these systems a promising coherent system." From this combination arises a new notion in computational area: cooperating objects.

In this sense, Embedded WiSeNts tries to detect the research trends about cooperating objects in the following 10 or 15 years, and to diffuse its results to the academic and industrial areas.

7.15 μSWn

- Name: Solving Major Problems in Microsensorial Wireless Networks
- Program: FP6 STREP Project # IST 034642
- URL: https://www.uswn.eu

Summary of the Project's Objectives

Focusing on the technological research side to obtain standard paradigms for unsolved problems related to WSNs, the μSWN project's main objective is to research generic and reusable software and hardware solutions that are common to existing and potential future applications. Moreover, the project focuses on researching and developing reusable middleware components to ease future development regarding similar systems under real-time restrictions. The research regarding the challenge of WSN architectural design and deployment will be focused on obtaining solutions that, although generic, allow for further fine-tuning when a given application is considered. Among the publications of this project done as of this writing, we would like to cite Stoyanova et al. (2007).

For the μSWN project, WSN technology has been applied to three specific application scenarios: surveillance, multi-tracking, and real-time event handling to be implemented at the Versmé Sanatorium in Birstonas, Lithuania. However, the advantages of wireless sensor technology can clearly go beyond these applications. Thus, solutions obtained in μSWN should be applicable not only to the three scenarios to be deployed as part of the project, but also to as wide a range of applications as possible.

References

Casaca A, Westhoff D (Eds) (2006) Scenario definition and initial threat analysis. UbiSec&-Sens Project Deliverable D0.1.

Havinga P, Etalle S, Karl H, et al. (2003) *EYES—Energy Efficient Sensor Networks*, Springer, Berlin.

Heinzelman WR, Kulik J, Balakrishnan H (1999) Adaptive protocols for information dissemination in wireless sensor networks. In *Proceedings of the 5th Annual ACM/IEEE International Conference on Mobile Computing and Networking (MobiCom '99)*, pp. 174–185.

Iftode L, Borcea C, Kang P (2004) Cooperative computing in sensor networks. In Ilyas M, Mahgoub I (Eds) *Handbook of Sensor Networks: Compact Wireless and Wired Sensing Systems*, CRC Press, Boca Raton, FL.

Intanagonwiwat C, Govindan R, Estrin D (2000) Directed diffu-sion: A scalable and robust communication paradigm for sensor networks. In *Proceedings of the 6th Annual International Conference on Mobile Computing and Networking (MobiCom '00)*, pp. 56–67.

IST EVENT 2006. Strategies for Leadership. http://ec.europa.eu/information_society/istevent/2006/index_en.htm.

Kang P, Borcea C, Xu G, et al. (2004) Smart Messages: A distributed computing platform for networks of embedded systems. *Comput J* 47:475–94.

Kulik J, Heinzelman W, Balakrishnan H (2002) Negotiation-based protocols for disseminating information in wireless sensor networks. *Wirel Netw* 8(2/3):169–85.

Disco Lab, Rutgers University. Smart Messages. http://discolab.rutgers.edu/sm/. Accessed October 2006.

Stoyanova T, Kerasiotis F, Prayati A, et al. (2007) Evaluation of impact factors on RSS accuracy for localization and tracking applications. In *Proceedings of the 5th ACM International Workshop on Mobility Management and Wireless Access (MobiWac '07)*, pp. 9–16.

Wireless World Initiative Home page. http://www.wireless-world-initiative.org/.

Zhou P, Nadeem T, Kang P, et al. (2004) EZCab: A cab booking application using short-range wireless communication. Rutgers University Technical Report DCS TR550.

Chapter 8
WSN Application Scenarios

Abstract Wireless sensor network (WSN) technology is a new and modern technology that has already been implemented in a wide variety of scenarios, and its applications are growing every day. As models for mobile wireless networking become more popular, their appeal comes from the fact that they can operate autonomously without the need for an existing infrastructure. This great benefit can be seen even more clearly when looking at the many problems that the use of WSN technology solves. The applications of WSN technology have been classified into four main categories: environmental monitoring, health care, security, and additional applications. After we review the different WSN cases in these fields, we identify three of the most demanding and most representative scenarios to demonstrate the advantages of WSN technology and its boundless potential in today's world. The technical analysis defines the use cases for multiple-target tracking, surveillance and vital sign, and environmental critical monitoring and specifies the user requirements and technical description of the network infrastructure and overall system that should be designed to support all three application scenarios.

8.1 Application Fields for WSNs

The applications of wireless sensor network technology have been broken down into four main categories: environmental monitoring, health care, security, and additional applications. The advantages of the technology can be clearly demonstrated through the WSN application classification even though, regardless of the application field, this technology has the capability to transform people's lives in all parts of society worldwide.

8.1.1 Environmental Monitoring

The use of wireless sensor networks in environmental applications is becoming more and more important in a world concerned about climate change, global

A.-B. García-Hernando et al., *Problem Solving for Wireless Sensor Networks*,
DOI: 10.1007/978-1-84800-203-6_8, © Springer-Verlag London Limited 2008

warming, and diminishing natural resources. WSNs can contribute to the development of hazard response systems, natural disaster detection systems, and energy-monitoring systems, among others.

Sensor networks have evolved from passive logging systems that require manual downloading, into intelligent sensor networks. These networks are comprised of nodes and communication systems that actively transmit their data to a sensor network server (SNS) where the data can be integrated with other environmental data sets. These sensor nodes can be fixed or mobile and range in scale depending on the environment being sensed. Large-scale single-function networks tend to use large single-purpose nodes to cover a wide geographical area. Localized multifunction sensor networks typically monitor a small area in detail, often with wireless ad hoc systems. Environmental monitoring can be broken down into five representative groups, as shown in Fig. 8.1: meteorological, geological, habitat, pollution, and energy monitoring.

8.1.1.1 Meteorological Monitoring

The main goal of meteorological monitoring is to control, supervise, and study several physical and atmospheric magnitudes. Traditional weather stations provide information about rainfall, wind speed and direction, air temperature, barometric pressure, relative humidity, and solar radiation. These measurements can be useful to forecast the weather and to predict or detect harsh natural phenomena (Bruno and Blumberg, 2006). The advantage of using WSNs in weather forecasting is the acquisition of large amounts of data that would be too difficult to obtain otherwise. This information can be stored in databases and then analyzed to improve the reliability of predictions. Natural meteorological disasters such as hurricanes, floods, and droughts have historically been the cause of great losses economically and in terms of human lives (Roark and Van Wie, 2003; Kung et al., 2006). In this context, the use of wireless sensor networks can be a way to reduce or prevent damage, where the benefits

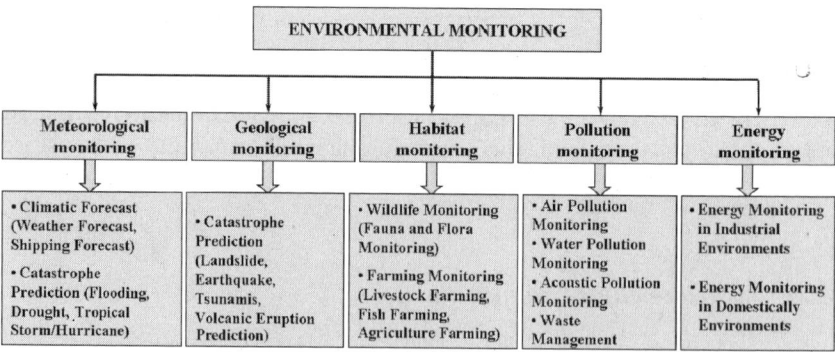

Fig. 8.1 Environmental monitoring classification

of this technology in foreseeing catastrophes are real-time data collection, coordinated and accurate response, and large monitoring areas. In the case of flooding, a WSN may provide real-time information on rainfall and water levels. In the case of drought forecasting, the network monitors and collects spatial and temporal ground surface information. Finally, for tropical storm and hurricane predictions, nodes can be deployed on the water surface, with water temperature, ambient humidity, and wind speed–sensing capabilities (Kung et al., 2006).

8.1.1.2 Geological Monitoring

Geological monitoring refers to the control, supervision, and study of several physical geological magnitudes, to enhance the understanding of the earth's state, with one of the most important applications being catastrophe prediction. The main feature shared by geological disasters such as earthquakes, tsunamis, volcanic eruptions, and landslides is the fact that they are related to an underground event (Evans et al., 2006). Unlike existing methods of monitoring underground conditions, which rely on buried sensors connected via wire to the surface, WSN devices are deployed completely belowground and do not require any wired connections.

8.1.1.3 Habitat Monitoring

Habitat monitoring refers to surveys aimed at detecting and explaining changes in the environment, both flora and fauna, and to assess the effects of any conservation action. Sensor network solutions for habitat monitoring show enormous potential benefits for the industrial and scientific communities, and society as a whole, because of their long-term data collection ability at scales and resolutions that are difficult to obtain otherwise through traditional instrumentation and their easy interaction with other external networks (Akyildiz and Stuntebeck, 2006; Biagioni and Bridges, 2002; Beckwith et al., 2004; Mainwaring et al., 2002). Researchers are becoming increasingly concerned about the potential impact of the human presence when monitoring plants and animals in field conditions, yet with wireless sensors, wildlife monitoring can be carried out without the use of traditional intrusive instrumentation.

8.1.1.4 Pollution Monitoring

A huge concern of the 21st century is the increase in pollution and its devastating effects. Air pollution, water pollution, noise pollution, and radioactive contamination, to name a few, need to be controlled and monitored in an attempt to reduce their damaging consequences. Classical systems have a straightforward and suitable way of centralizing information coming from sensors monitoring air pollution: the use of cables. However, WSNs have introduced a new way of operating, in which the nodes are organized into an ad hoc wireless architecture (Baumgartner and Robert, 2006). In the example of a water pollution detection system in a lake located near a factory that uses chemical substances, sensor nodes can be randomly deployed in unknown and hostile areas and relay the exact origin of a pollutant to authorities, who can then take appropriate measures to limit the pollution's spread.

8.1.1.5 Energy Monitoring

The cost of energy has become a significant factor in the performance of economies and in the maintenance of the environment, making energy resource management of utmost importance. The production and consumption of energy resources is very important to the global economy, and often it is possible to save on expenditures by incorporating innovative technology and new management techniques. The advantage of using wireless technology is that the energy waste can often be reduced by something as easy as measuring the temperature or human presence in a room and taking the necessary steps such as switching off a light or turning down the heat (Sensicast, 2008).

8.1.2 Health Care

WSN technology could potentially impact a number of health-care applications, such as medical treatment, pre- and post-hospital patient monitoring, people rescue, and early disease warning systems. In addition, WSNs can contribute to solving some important social problems, such as caretaking of the chronically ill, of the elderly, and of people with mental and physical disabilities. This will not only improve these people's quality of life, but also benefit society as a whole.

Since the health-care domain is a very broad grouping, it has been divided into five separate categories as presented in Fig. 8.2: patient monitoring, disability assistance, people rescue, bio-surveillance, and smart surrounding.

8.1.2.1 Patient Monitoring

The main goal of patient monitoring is to observe the patient's state of health in the hospital and/or home environments. Current systems used for long-term patient monitoring require the use of wires, whereas in WSNs this is not necessary. The measurements of patient vital signs can be useful not only for medical records and treatments, but also for later rehabilitation. Patient monitoring in hospital environments aims at continuously collecting patients' vital signs. In this scenario, every patient could wear tiny, wireless vital sign sensors, allowing doctors and nurses to continuously monitor their status and to react to changes (Lo et al., 2005; HEARTS).

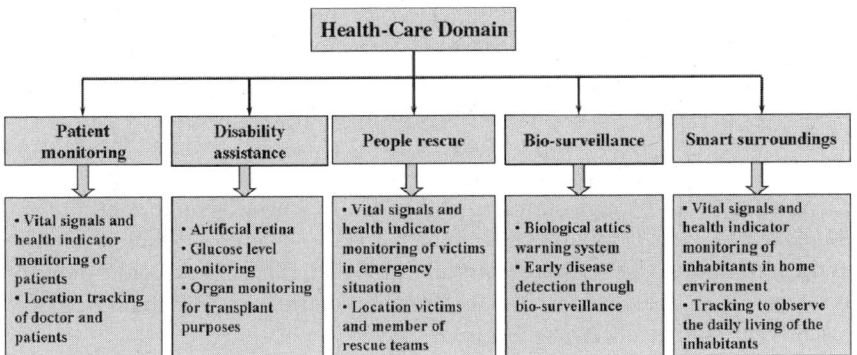

Fig. 8.2 Classification in the health-care domain

8.1.2.2 Disability Assistance

Disability assistance considers application scenarios where smart sensors operate within the human body to counteract organ weaknesses or to monitor important physiological parameters or particular organ viability. For example, when managing patients with diabetes, the blood glucose level can be monitored continuously, controlling the insulin delivery from an implanted reservoir (IBM, 2005). For the treatment of epilepsy and other debilitating neurological disorders, implantable, multi-programmable brain stimulators are already on the market, which save the patient from surgical operations such as removing brain tissue. In cardiology, the value of the implantable cardioverter-defibrillator has increasingly been recognized for the effective prevention of sudden cardiac death (Van Laerhoven et al., 2004). Such technological development reflects the social, industrial, and clinical perspectives of future health-care delivery.

8.1.2.3 People Rescue

The advances in wireless networking and medical sensors extend the possibilities for providing emergency care. In emergency or disaster scenarios, if people are outfitted with tiny wireless badges, rescue teams and medics will be guided much faster to victims, allowing large numbers of casualties to be prevented (Gao et al., 2005). These sensors would relay continuous data to nearby paramedics and emergency medical technicians, who would use mobile PDAs or mobile PC-based systems in ambulances to capture all vital patient data. They could thus monitor and care for several patients at once and be alerted to any changes in the patient's physiological status. The information network includes communication with the rescue teams as well as with the hospitals information system, allowing for better coordination between the emergency rescue teams and the hospitals with the facilities and resources to care for patients in critical condition. Patients in accidents can greatly benefit from technologies that continuously monitor their vital status and track their locations until they are admitted to the hospital (MobiHealth). For example, the first rescuer who arrives at the area of an accident with a large number of victims would place wireless vital sign and location-tracking sensors on each patient. The sensors provide several functions: vital sign monitoring, location tracking, medical record storage, and triage status tracking. These sensors would continuously relay data to nearby paramedics and emergency medical technicians, who would use mobile PDAs to capture all vital data. Thus, they could monitor and care for several patients at once and be alerted to any changes in the patient's physiological status.

8.1.2.4 Bio-surveillance

All wireless sensor systems created with the purpose of bio-surveillance help public health experts determine the likelihood of a deadly disease outbreak among humans (Shea and Lister, 2003). A series of sensors can collect and examine samples from the air, soil, and water and use weather conditions to predict the epidemiological spread of the disease (Anderson et al., 2003). This prediction allows federal, state, and local officials to react, providing fast emergency response, medical care, and subsequent management needs.

8.1.2.5 Smart Surrounding

Smart sensor technology can help solve some important social problems, such as caretaking for the chronically ill, the elderly, and people with mental and physical disabilities. Often this high level of health care can be provided in citizens' homes while they continue a normal, active life instead of being institutionalized (Medical Home; Sanders, 2000). This will not only improve an individual's quality of life, but will also be a great benefit to society as a whole.

8.1.3 Security Domain

This application domain not only includes a wide variety of challenging problems like target tracking and localization, detection of toxic chemicals, rescue, and homeland security, but has also defined the common view of a wireless sensor network as a large-scale, multi-hop ad hoc network of tiny resources and energy-constrained sensor nodes. In addition, the cost involved in using a sensor network for military applications is of less importance if technology can provide a strategic advantage in warfare (DARPA, 1997).

European research in WSNs focuses on civilian applications. In this context, we have decided to map military applications to the civilian domain; for instance, using material flow monitoring for the logistics of dangerous goods and using tracking in building security. Sensor networks for these applications may have different properties, such as being carried out on a smaller scale, heterogeneous hardware, single-hop networks. Typically, it is hard to convince potential users of these applications to invest money in research, at least when compared to military funding.

Surveillance is taken as the process of monitoring the behavior of people, objects, or processes within systems, for security or social control. WSN technology is very well suited for surveillance systems, mainly because wireless sensor networks do not require any wired infrastructure. However, surveillance needs to be categorized in order to take into account the different requirements of the different cases, namely indoor and outdoor surveillance, as shown in Fig. 8.3.

8.1.3.1 Indoor Surveillance

Indoor surveillance has two possible application scenarios: surveillance systems placed in a private environment such as a home, and those located in public buildings such as hospitals, museums, and airports, among others. Wireless indoor sensor networks differ from general sensor networks in that they can have nodes with heterogeneous resources and at the same time dissimilar attributes. The nodes may have different energy capacities, processing capabilities, positions, and radio coverage. Wireless networks have many advantages over their wired counterparts that need to be taken into consideration as well. They are easily deployed, have ubiquitous connection, are low in maintenance, and are unobtrusive.

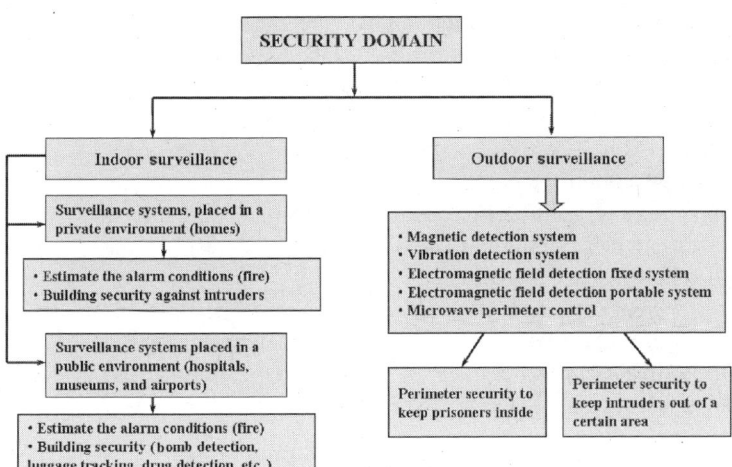

Fig. 8.3 Security domain classification

WSNs offer new possibilities in public building surveillance, such as saving costs in wiring installation. Some sensors, such as thermal sensors, integrate processing units that estimate alarm conditions; thus, presence detection can be carried out with volumetric sensors. These sensors receive infrared radiation from elements or generate an invisible wide detection field that is hard to avoid. Like electrostatic sensors, most volumetric devices are based on pyroelectric sensors, which convert infrared radiation variations emitted by elements into a small current, allowing this sensor to examine infrared radiation in a fixed area of the room, where it is installed. Video surveillance systems are necessary to identify the alarm source detected by volumetric devices; a small amplitude and speed variation at any given time will trigger the alarm. Nevertheless, a combination of both devices is possible. With this purpose, movement video detection systems transform the video-capturing capabilities of a closed television circuit into an image-capturing detection system, analyzing video output to generate an evaluation field. The main advantage of these systems is that no additional video surveillance system is required, since video monitoring offers information about the alarm's source, helping to minimize false alarms (DARPA, 1997).

8.1.3.2 Outdoor Surveillance

Outdoor surveillance is also highly important for perimeter security, such as keeping prisoners inside the premises or keeping intruders out of a certain area (Chang et al., 2004). When using invisible surveillance, it is fundamental that intruders are not able to detect its presence and then sabotage the detection system. For this reason, several technologies have been developed to yield a cost-effective solution for particular proprietors.

- *A magnetic detection system:* This method is based on the magnetic anomaly's passive detection, which allows the detection of any intruder that carries ferromagnetic metal objects.
- *A vibration detection system:* Vibration sensors are attached to a wire fence, which detect any vibration produced by climbing or wire-cutting of the enclosure (Magal, 2002).
- *An electromagnetic field detection fixed system:* This technology makes use of a fixed set of volumetric sensors for perimeter intruder detection, which generates an electromagnetic field around two wires, both field-emitter and receiver, buried in the ground.
- *An electromagnetic field detection portable system:* Portable systems allow fast installation and easy handling for perimeter security. ·
- *A microwave perimeter control:* A microwave barrier is established between sensors, whose status is analyzed by a digital signal processor.

These systems are based on amplitude and phase disturbance detection and use a small antenna to emit the detection signal. When intrusion activity patterns are received in the perimeter antenna, the system activates an alarm.

These systems can be combined with WSN technologies, as each antenna forms an isolated node.

Airport security provides a first line of defense by attempting to stop potential attackers from bringing weapons or bombs into the airport. WSN technology plays an important role here. If airport security is successful, then the chances of these devices getting onto aircrafts are greatly reduced. Bomb detection, luggage tracking, drug detection, and hijacking are all current threats in many airports throughout the world, which is why the advancements in WSN technology are so important today.

A nation's military forces are extremely important. Military defense requires constant industrial and technological advancements. The benefits of WSNs can be seen in homeland security, military vehicle operation, and maintenance and battlefield monitoring. One of the most important in today's world is the application of this technology in homeland security, where intrusion detection and perimeter monitoring are key elements to successful homeland defense in this new age of terrorism. Active and passive sensors can be used to detect the presence of nuclear, biological, and chemical agents. Passive sensors detect a change in the natural energy field caused or emitted by a target. Passive sensor technology includes those sensors based on capacitance, heat, sound, and vibration. Active sensors transmit energy and detect a change in the received energy as the target comes within range. Local law enforcement plays another important role in homeland defense. Advanced wireless networks can aid police departments in tracking down suspects and responding to criminal activity. By linking wireless routers to video cameras, images can be transmitted from disaster areas to vehicles, fire stations, command centers, and other public safety agencies. Officers in command centers can view the video of an incident, analyze footage, and relay orders to a response team, all in real time. The detection of biological, chemical, and nuclear attacks and sniper localization can all make important use of wireless sensor technology.

8.1.4 Additional Domains

The previous sections presented three important specific application scenario domains. We can expand the scope of the scenarios by asking, "How can a WSN be applied to other domains or fields?" A complete list of applications is limited only by the imagination, but in order to answer this question, we will focus on several applications that have already been mentioned by users and companies. Since WSN technology can be applied in such a wide variety of ways, this list of additional applications will grow as potential users become aware of the technological capabilities. Today, many of the WSN applications described in this section are some of the most widespread. Depending on the needs of the company or user, this grouping of applications could be seen as one of the most important. The list of applications includes areas like structural health monitoring, building monitoring and control, automotive monitoring, traffic monitoring, industrial process control, and asset and warehouse monitoring.

8.1.4.1 Structural Health Monitoring

Life-cycle monitoring of civil infrastructures such as bridges and buildings is critical to the long-term operational cost and safety of aging structures. It is in this context that WSNs are receiving special attention in an attempt to minimize cost and maximize the utility of the system as a whole by performing real-time monitoring (Sazonov et al., 2004; Pakzad et al., 2005). Events such as earthquakes can cause enormous damage to civil infrastructures without producing any apparent visible damage. Such damage can result in life-threatening conditions in the structure either in the immediate aftermath or long after the actual event has occurred. Near real-time structural monitoring of civil infrastructure reduces the loss of human lives by warning the appropriate authorities about hazardous structures and impending collapses and provides information to emergency response services. For bridge structure monitoring in particular, various techniques can be used:

- *Slow monitoring* measures slow phenomena like temperature changes, settling, and concrete relaxing. Therefore, the sampling frequency can be hours. Variables measured could be air temperature, straining in several axes, steel distortion, and solar radiation, in places like boards and pillars.
- *Fast monitoring* measures fast phenomena like traffic, wind, and earthquake effects. Sampling frequency may be variable, seconds, or milliseconds, with levels depending on variable speed changes. The sensors for these measurements are accelerometers, wind speed, and direction meters.
- *Corrosion monitoring* is designed to detect steel corrosion and thus has a slow measurement interval of perhaps days.

8.1.4.2 Building Monitoring and Control

Sensors embedded in a building can drastically decrease energy costs by monitoring the building's temperature and lighting conditions. The information obtained is then used to regulate heating systems, cooling systems, ventilators, lights, and computer servers (Zhao and Guibas, 2004). If a conference room full of people becomes too hot, cold air may be borrowed from an adjacent room that is temporarily empty for the next couple of hours. Sensors in a ventilation system may also be able to detect biological agents or chemical pollutants. Wireless sensors are also an attractive alternative to wire-controlled devices such as light switches because of the high cost of wiring. Wireless sensor networks are easy to install, making this solution a suitable technology to monitor a wide variety of energy measurements and receive automatic notification upon detection of unusual events.

8.1.4.3 Automotive Monitoring

Automotive monitoring and traffic and transport monitoring are two very important applications of wireless sensor technology. In both cases, the physical quantity that is usually used to measure is the magnetic field (Crossbow, 2007), which in turn conditions the technology that will be used to measure the magnetic field. The sensor module detects passing vehicles by measuring disturbances in the earth's magnetic field caused by passing vehicles. Almost all of today's road vehicles, even vehicles with polymer body panels, contain a large mass of steel. The steel has a much higher magnetic permeability than the surrounding air, which concentrates the flux lines of the earth's magnetic field, increasing the magnitude of the B-field inside and in the immediate vicinity of the vehicle. This disturbance is detectable as far away as 15 meters from the vehicle.

8.1.4.4 Traffic Monitoring

Increasing congestion in public road networks is a growing problem in many countries (Coleri et al., 2004). Any remedial strategy for the efficient management of roads requires the measurement of traffic conditions. For instance, the traffic management center (TMC) uses measurements of traffic at urban intersections to optimize traffic signal settings based on traffic queue lengths. Road users can use this information to better plan their activities and adjust their routes. Most conventional traffic surveillance systems use intrusive sensors, including inductive loop detectors, microloop probes, and pneumatic road tubes, because of their high accuracy for vehicle detection ($>97\%$). However, these sensors disrupt traffic during installation and repair, which leads to a high installation and maintenance cost. Wireless sensor networks and access points can be used so that traffic information is generated at the sensor nodes and then transferred to the access point by radio.

8.1.4.5 Industrial Process Control

The value of wireless networks is becoming obvious to organizations that need real-time access to information about the environment of their plants, processes, and equipment to prevent disruption (Conant, 2006). Wireless solutions can offer lower system, infrastructure, and operating costs as well as improvement of product quality, streamlining of operations, easier upgrading, greater physical mobility, and more freedom. Unlike traditional wired networks, the sensors in a WSN can be deployed in the bearings of motors, oil pumps, whirring engines, packing crates, and many other unpleasant, inaccessible, or hazardous environments that are inaccessible with normal wired systems (Low et al., 2005). By using smart sensors, the condition of equipment in the field and factories can be monitored to alert the appropriate authorities of imminent failures. A typical equipment manufacturer spends billions of dollars on service and maintenance every year. The equipment to be monitored can range from turbine engines to automobiles, photocopiers, and washing machines. Condition-based monitoring is expected to significantly reduce the cost of service and maintenance, increase the machine's lifetime, improve customer satisfaction, and even save lives.

8.1.4.6 Asset and Warehouse Monitoring

Sensors may be used to monitor and track assets such as trucks or other equipment, especially in an area without a fixed networking infrastructure (Zhao and Guibas, 2004). Sensors may also be used to manage assets for industries such as oil, gas, and aerospace. Tracking sensors can vary from GPS-equipped locators to passive RFID tags, and the automated logging system can reduce errors in manual data entry. More importantly, businesses such as trucking or construction can significantly improve asset utilization using real-time information about equipment location and condition. Furthermore, the asset information can be linked with other databases such as Enterprise Resource Planning (ERP) databases. This information helps to make decisions by giving a global, real-time picture in order to optimally use available resources.

8.2 The Three Most Prevailing WSN Application Scenarios

Three application scenarios are chosen to demonstrate the potential of WSNs for monitoring people's health and location and for ensuring perimeter security. We use the Versmé sanatorium, located in Birstonas, Lithuania, to demonstrate the three scenarios, as shown in Fig. 8.4. Around the artificial ponds, an invisible barrier with a 5-meter perimeter is assumed, while the distance from the building walls to the edge of the ponds is up to 20 meters. Fixed sensor nodes are placed on the trees in order to carry out environmental protection services as well. Other possible positions for the fixed sensor nodes are the on lampposts around the perimeter of the area and on the roof of the building. The building is 15 meters high, while the height of the trees can be up to 10 meters, and a power supply is available within the specified perimeter.

After reviewing a large variety of possible WSN application scenarios, three of the most demanding, and thus most interesting as far as research is concerned, have been selected. These three application scenarios have been chosen as the most representative of existing and future WSN applications with universal usage in the environmental, health-care, and security domains.

- *Application scenario 1, "Multiple-target tracking":* The WSN system targets moving objects inside the limits of the sanatorium, locates them, stores historical data, and obtains statistics regarding the preferred routes of clients.
- *Application scenario 2, "Surveillance":* The WSN system creates a virtual security perimeter to keep intruders out of the sanatorium.

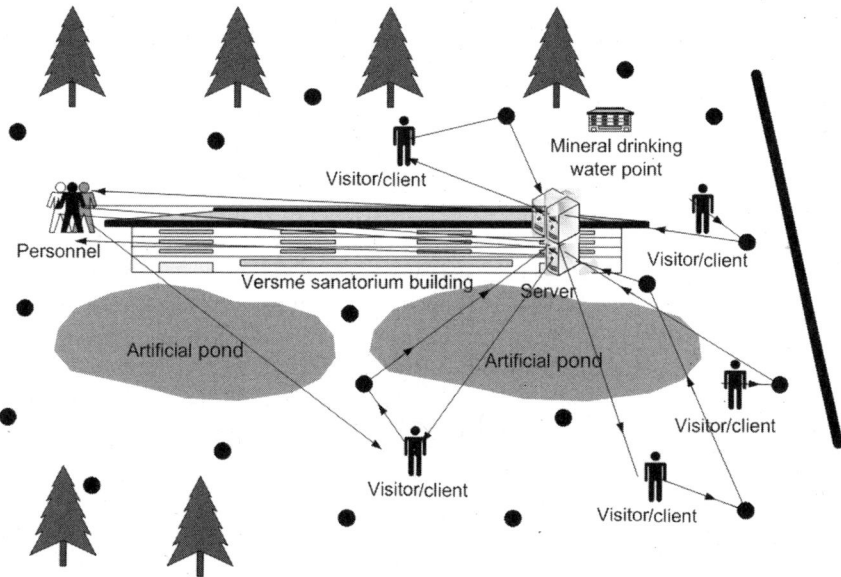

Fig. 8.4 Application scenarios: multiple-target tracking, surveillance, and vital sign and environmental critical monitoring

- *Application scenario 3, "Vital sign and environmental critical monitoring"*:
  The WSN system monitors vital signs in real time, keeping track of clients' health.

The three scenarios will be further described, corresponding to the aforementioned application fields, with respect to the scenario description from the user perspective and the resulting user requirements and specification. Last but not least, according to the formal object-oriented design approach (Pilone, 2006), technical requirements for each application scenario are derived from the identified use case analysis. Technical requirements are derived from the technical aspects that the system must fulfill, such as performance-related issues, reliability, and system availability. These types of requirements can be conceptually categorized in the following groups: functional, storage, interface, usability, reliability, communication, performance, maintenance, and documentation.

8.2.1 Multiple-Target Tracking

8.2.1.1 Scenario Description

Tracking can be defined as the pursuit of moving objects. From a technical point of view, it is a challenge to track targets without the use of global positioning system (GPS) technology, yet there are applications in which GPS signals cannot be accessed. However, WSN technology can be used to track targets, providing an alternative to GPS. For target tracking, the system localizes and tracks moving objects inside the limits of a specified area, to store historical data and obtain statistics regarding the preferred routes of visitors, residents, or patients, known as clients. The system also provides the staff with the precise position of clients and employees, which is vital information in emergencies. The system also provides suggestions about abnormal situations, comparing a client's historical data with current data to establish if the client's behavioral pattern makes sense. If the system detects significant changes in the client, the staff is notified and can then use their judgment to react to the warnings.

This scenario is implemented outdoors, in an area including a park, artificial ponds, and the sanatorium building. The total area outdoors covers 72.000 square meters, 11,000 square meters of which are covered by the two ponds.

A group of clients is strolling through the sanatorium's gardens. Clients wear smart bracelets that send positional information to a WSN system, which in turn processes the received data to determine the location of moving clients and staff. The staff carries PDAs or mobile phones with a user-friendly application to track the position of the clients staying at the sanatorium so that the personnel can quickly locate clients when needed.

Positional data will also be stored by the system and can be analyzed to reveal if clients prefer to walk a certain route around the sanatorium. Using this information, the sanatorium manager can decide to build a café on this route for the clients to have a drink or snack during their walks.

During his or her daily walk around the pond, a client can be tracked by the system. For example, if the client falls, the system detects that the client has not moved for a period of time. The system compares this data with historical data and can judge if the client is far from his or her usual route. The received data also indicate whether anyone is near the client. The system can establish if this is a potentially abnormal situation, thus sending a warning to the PDA or mobile phone of the staff member closest to the client, who can then check on the client to see if he or she needs help or possible treatment.

Apart from the health spa scenario, there are many other scenarios in which tracking can be useful. Multiple-target tracking applications supported by wireless sensor networks include surveillance, search and rescue, disaster response, pursuit evasion games, distributed control, spatial temporal data collection, health monitoring, person and group monitoring, and other location-based services.

8.2.1.2 User Requirements

The main goals of the multiple-target tracking demonstration scenario are to (1) measure data needed to track moving objects, (2) process the data collected, (3) analyze the results, and (4) provide suggestions to the sanatorium personnel. Clients wear smart sensor nodes that send positional information to a WSN. The WSN system processes the received data to determine the location of moving clients and personnel, with an accuracy of 10–12 meters. Fixed sensor nodes deal with more precise positional information that must be sent in real time, allowing the system to determine the position of moving clients.

The personnel have PDAs or mobile phones with an application that is designed to be very user-friendly. Personnel using the application on their PDAs or mobile phones are able to track the position of clients in the sanatorium. This positional data will also be stored by the system. The positional data can be analyzed to reveal that clients prefer to walk a certain route around the sanatorium. Using this information, the sanatorium system administrator could make some specific decisions to improve the environment for clients walking outdoors.

Measured data are transmitted to the system and processed to determine the geographical location of the moving objects. All the data results are stored, and sanatorium personnel can use this information. Long-term studies can define the patterns followed by clients, allowing the construction of different facilities around the routes or making better routes to attract more clients.

Table 8.1 shows the preferred features for the target-tracking application scenario.

8.2.1.3 Technical Analysis

This section identifies and analyzes the use cases and corresponding actors of the multiple-target tracking application scenario. Based on this information, the technical requirements for the final system are extracted.

Table 8.1 User Requirements for Multiple-Target Tracking

Parameter	Preferred Features
Mobility	Fixed beacon sensor nodes and mobility wearable sensor nodes
Energy resources	Rechargeable mobile nodes and battery-powered fixed nodes
Deployment	Manual preliminary predicted node placement
Heterogeneity	Body sensors, PDAs, data receiver, and specialized sensors
Scalability	Medium number of possible future nodes
Topology	Star, mesh, multi-hop, single-hop, hierarchical cluster-based
Coverage	Full area coverage
Connectivity	Full network connection
Communication modalities	RF-based
Infrastructure	WANs, monitoring nodes, cluster members, cluster heads, gateways, base station, PDAs
Sensor cost	Low-cost*
Sensor type	Tiny, nonintrusive, easy to deploy, easy to wear, robust
Network size	Up to hundreds of fixed nodes and tens to hundreds of mobile nodes
Network lifetime	Months
Application-dependent QoS parameters	Data reliability, time delay (real-time needs), data privacy and security, low end-to-end delay, time synchronization, fault tolerance

*Sensor nodes cost from 100€ to 400€ each.

Actors

Client: represents the sensor nodes worn by the clients. The sensor node will set off an alarm in an emergency and will provide the client's information to be stored in the central database.

Event scheduler: represents the main component that handles actions received as events, automatically schedules them, and coordinates the functional components of the system. This actor is responsible for setting off alarms, localizing and keeping track of moving clients, checking the status of the nodes, and processing historical data.

System administrator: represents the system manager in charge of the system, with additional access rights in order to process clients' personal data or, if he or she is a doctor, with access rights to clients' historical health data. The system administrator may also process statistics about clients' routes, register and unregister new clients, manage their information, and present stored data from the database.

Personnel: represents the staff working in the sanatorium, who may request the position of a client, recall stored data, and confirm alarm notifications upon reception.

Use Cases

Node status checking: The event scheduler may periodically check the status of the wireless nodes to identify failures and notify the system administrator.

Register/unregister new client: For any new client, the system administrator registers the relative information in the system database, forming a corresponding record for later use. The system administrator has the right to unregister the corresponding record of any registered client and remove it from the system database.

Manage client information: The system administrator may modify the stored information in its corresponding record for any registered client in the system database.

Client localization: Registered clients continuously send information to the system that allows their position to be calculated. The system processes the received data and executes a tracking algorithm to determine the location of moving clients and personnel. Fixed sensor nodes deployed in the sanatorium area deal with positional information. The positional data must be sent in real time in order to guarantee that the system is able to determine the position of moving clients.

Store registered client information: The positional data sent by the fixed sensor nodes in real time is used to determine the position of moving clients. The resulting positional client information will be routed to the central server for final storage in the central database.

Statistical processing of routes: The stored positional and historical data can be analyzed statistically to reveal that clients prefer to walk a certain route around the sanatorium. This information may be used, for instance, to support the system administrator's decision to build a bench on this route where the clients can rest during their walks. Additionally, the event scheduler may periodically process statistical data to identify emergency situations and notify the personnel, according to the defined alarm rules.

Historical data processing: The event scheduler may periodically process historical data to identify emergency situations and notify the personnel.

Stored data presentation: The stored historical data can be recalled for specific periods of time, for specific clients, or for groups of clients to support system-wide managerial decisions regarding the preferred services of the sanatorium.

Client position request: The personnel may request the position of a client for a daily check. Additionally, a client who is lost or in any other emergency situation may request help using an alarm button, thereby revealing his or her position.

Indicate alarm situation: The system continuously receives positional data, processes it, and compares it with stored data. When the system detects an abnormal situation, it analyzes whether any help close by can be provided to the localized emergency and sends an alarm notification to the personnel closest to the problem.

Confirm alarm notification: In an abnormal situation, the system is responsible for sending an alarm notification to the personnel. If the staff member can handle the emergency, he or she sends an affirmative message. Otherwise, if the system receives a negative response or no response, then it notifies the next-closest personnel of the emergency.

8.2.2 Surveillance

8.2.2.1 Scenario Description

The use of WSNs can greatly improve security in a close area and thus create a safer environment. Suspicious situations can also be detected; thus, potentially dangerous situations can be avoided. The surveillance system creates an invisible security perimeter that detects objects crossing it in order to keep intruders out of the sanatorium. The system also stores historical data and obtains statistical information regarding the entrances and exits of clients, staff, and possible intruders. The system provides staff with suggestions on how to improve security in the sanatorium and also how to take special care of clients who must stay within the boundaries of the facility.

A WSN is used to detect objects crossing the invisible barrier around the perimeter of the sanatorium. Clients and staff wear smart bracelets for their identification. If an intruder (person or animal) enters the limits of the sanatorium, the movement is detected by the WSN, which sends the information to the central system. The system analyzes the data received, decides whether or not the person or animal that has crossed the invisible barrier is an intruder, and sends an alarm notification to the appropriate personnel.

When a client enters the limits of the sanatorium after a walk outside, the movement sensors detect him or her and the central computer determines if he or she is a client according to the smart signal that has been received. Each client or staff member wears an identifying smart bracelet. As a result, the central computer knows who crosses the sanatorium's perimeter. Consequently, the alarm is obviously not activated when staff members or clients cross the Parameter, thus saving system power.

A variation of this scenario entails a client exiting the limits of the sanatorium. The movement is detected and information is sent to the central computer. Each client has a profile stored in the database that indicates if he or she is allowed to exit the limits of the sanatorium without supervision. Once the stored profile of the client is processed, the system identifies a potentially dangerous situation and sends an alarm to the PDA of the responsible staff. The staff could then decide to search for him or her in the sanatorium's surroundings.

8.2.2.2 User Requirements

The main goals of the perimeter surveillance scenario are (1) the detection of intruders, (2) the processing of information regarding detection, (3) the analysis of the results of that data, and (4) the creation of a self-learning system that improves its performance with time. Institution territory surveillance involves

- Detection of intruders and all the visitors, patients, and staff passing through the territory
- Monitoring the activity in the sanatorium's facilities including therapy rooms, mud bath, swimming pool, and resting room
- Automatic monitoring and regulation of temperature, moisture, and lighting on the premises

The system creates a 31,000-square-meter virtual security perimeter that detects objects crossing the invisible perimeter in order to keep intruders out of the sanatorium. The system also stores historical data and obtains statistics regarding the entrances and exits of clients, personnel, and possible intruders. The system provides personnel with suggestions to improve security in the sanatorium and also to take special care of clients who must stay within the facility's boundaries.

The perimeter is constantly observed in order to detect intruders at once. An informational signal (light or text) is generated when crossing a 5-meter-long invisible line from the edge of the perimeter. An alert message must be received by personnel closest to the client. The sensor nodes should distinguish among clients, personnel, and intruders. To achieve this objective, clients and personnel wear sensor nodes to send a signal identifying clients and personnel. If no signal is received from a sensor node, an additional type of sensor node is

Table 8.2 Surveillance Application's User Requirements

Parameter	Preferred Features
Mobility	Fixed beacon sensor nodes and mobility wearable sensor nodes
Energy resources	Rechargeable mobile nodes and battery-powered fixed nodes
Deployment	Manual placement
Heterogeneity	Emplaced sensors, PDAs, data receiver, specialized sensors, infrastructure sensors
Scalability	Medium number of possible future nodes (+100s)
Topology	Star, mesh, multi-hop, single-hop, hierarchical cluster-based
Coverage	Full area coverage
Connectivity	Full network connection
Communications modalities	RF-based
Infrastructure	WANs, fixed and mobile nodes, cluster members, cluster heads, gateways, base station, PDAs
Sensor cost	Low-cost*
Sensor type	Tiny, nonintrusive, easy to deploy, easy to wear, robust
Network size	Tens to hundreds of fixed and mobile nodes
Network lifetime	Months
Application-dependent QoS parameters	Data reliability, time delay (real time needs), data privacy and security, fault tolerance, low end-to-end delay, time synchronization

*Sensor nodes cost from 100€ to 400€ each.

activated to gather further information necessary for the system to determine that an intruder is crossing the perimeter. A combination of different types of sensor nodes is thus necessary.

When making decisions, the system has to take into consideration all the information it receives from the sensor nodes. Historical data in addition to client profiles stored in the database can be used to detect potentially abnormal situations and make suggestions to personnel by activating an alarm in his or her PDA.

Table 8.2 shows the preferred features for the surveillance application scenario.

8.2.2.3 Technical Analysis

This section identifies and analyzes the use cases and corresponding actors of the surveillance application scenario. Based on this information, the technical requirements for the final system are extracted.

Actors

Client: represents the sensor nodes placed along the surveyed perimeter. The sensor node will be triggered to detect when an intruder crosses the perimeter.

Event scheduler: This actor represents the main component that handles actions received as events, automatically schedules them, and coordinates the functional components of the system. As in the multiple-target tracking scenario analysis, the event scheduler is the actor responsible for setting off alarms, checking the status of a sensor node, and processing historical data for predicting emergencies.

System administrator: represents the system manager in charge of the system, with additional access rights in order to process clients' personal data or, if he or she is a doctor, with access rights to clients' historical health data. The system administrator may also process statistical information on clients' entrances and exits, register and unregister new clients, manage their information, and recall stored data from the database.

Personnel: represents the staff working in the sanatorium, who can recall stored data and confirm alarm notifications upon reception.

Use Cases

Crossing of perimeter detection: The surveyed perimeter is formed around the sanatorium area (31,000 square meters) by appropriately deployed sensor nodes to deal with intrusion detection. When a moving object, be it an intruder, client, or staff member, crosses the imaginary security perimeter created by the wireless sensor network, the system detects the object crossing the perimeter using a combination of different types of sensor nodes. The detection must be done in real time.

Moving object identification: The fixed sensor nodes are responsible for dealing with intruder identification. These nodes are placed outdoors and should distinguish among clients, personnel, and intruders. To achieve this objective, clients and personnel are assumed to be wearing sensor nodes, which send an identification signal so they can be considered as registered. As a result, the network of sensor nodes alerts the base station when people cross the sanatorium's perimeter.

Alarm indication: The system receives the location of the people crossing the security perimeter and stores it for future processing. If an intruder is inside the perimeter or a client is outside the perimeter, the system detects it based on the stored data and sends an alarm notification to the personnel closest to the localized emergency.

Process historical data, entrance and exit statistics: The stored historical data can be recalled for specific periods of time, for specific clients, or for groups of clients to support managerial decisions regarding the sanatorium's security. Moreover, the stored historical data regarding the entrances and exits can be analyzed statistically to reveal the weakest or potentially dangerous spots in the security perimeter. Using this information, the system administrator can decide to build a fence or assign more personnel to that region. Additionally, the event scheduler may periodically process historical data to generate behavior patterns and thus identify emergency situations.

8.2.3 Vital Sign and Environmental Parameters

8.2.3.1 Scenario Description

Monitoring can be defined as the repeated observation of conditions, especially to detect and give warning of change. The need for data monitoring under critical conditions as well as prioritization of emergency notifications is the main challenge of this scenario. Critical monitoring is considered a failure if it is not completed before a specific time threshold. Therefore, time limits or deadlines will have to be clearly defined relative to the specific events being monitored.

The goal of vital sign and environmental critical monitoring is to monitor vital signs to keep track of the health of clients, to store historical data, and to obtain statistics regarding the vital signs of clients at the sanatorium. The system also monitors the environmental Parameters of the sanatorium and provides suggestions regarding an increase or decrease in the temperature and/or humidity in any area of the building in order to create a comfortable environment for all guests. Specialists will be able to recommend treatments to clients based on what their vital signs indicate. They will also be able to determine if and when there are health emergencies and what the reasons are for those emergencies.

This scenario assumes that clients of the sanatorium wear smart sensor nodes to detect his or her vital signs and send these data to a WSN. The smart sensor node also contains an alarm indication mechanism to inform clients of potentially dangerous situations. There is a network of fixed sensor nodes inside and outside the building. The outdoors territory is 31,000 square meters.

One example of the use of WSN technology in client vital sign monitoring occurs when a client participates in a remedial gymnastics class in the knees therapy hall (the area of coverage is 281.65 square meters). The smart sensor node the client is wearing sends vital sign information every minute while the client exercises. The WSN system receives and processes this monitoring information. The system compares historical data for this client with current data and establishes that his or her pulse rate is abnormal and higher than the maximum expected pulse rate for this person. The system then sends a warning to sanatorium personnel via their PDAs or mobile phones and to the client via his or her wearable sensor node. The alarm is displayed on the personnel's PDA in the form of a warning, which suggests informing the client to stop taking the class. Their smart sensor node starts to beep. The client knows that when the smart sensor node beeps, he or she needs to see a specialist immediately. If the client decides to ignore the alarm and continue taking the remedial gymnastics class, his or her pulse rate may suddenly increase sharply. The system would then alert the personnel on their PDA devices or mobile phones that the client may be about to suffer a heart attack.

In order to monitor environmental data, sensors must be deployed to sense the temperature and humidity in all areas of the sanatorium. The personnel can then decide to increase or reduce the temperature and/or humidity of a particular area accordingly. Data about environmental Parameters can also be used so that other clients can avoid uncomfortable conditions inside the sanatorium, such as extreme heat or cold. Additionally, the system can be used for cross-referencing searches of historical data, to find out if the building's temperature has influenced changes in pulse rate.

8.2.3.2 User Requirements

The main goals of the critical monitoring scenario are to (1) measure pulse rate and environmental Parameters such as temperature and humidity, (2) process the data collected, (3) analyze the results, and (4) provide suggestions to sanatorium personnel. Two kinds of Parameters are measured in this application: vital signs and environmental data.

The vital sign that will be analyzed in this scenario is pulse rate, which measures heartbeats and other potential coronary responses. In order to avoid heart problems, among other reasons, this vital parameter has to be measured continuously while a client exercises. This application also measures environmental parameters, i.e., temperature and humidity. The measurement of environmental parameters may help to analyze the relationship or correlation between changes in these environmental Parameters and changes related to heart response.

The client wears a smart sensor nodes that continuously detects his or her vital signs and sends these data to the WSN system every minute the client exercises. The system receives this monitoring information and compares historical data for this client with current data to establish whether or not his or her pulse rate is abnormal. It is essential that the system is capable of guaranteeing that the vital signs are processed in real time so that emergency actions can be taken immediately. The alarm notification sent to the client's sensor node stops after 10 seconds and starts again if the problem persists. If an emergency is detected early, in cases such as heart attacks, the client's life may be saved.

In order to observe an individual's pulse rate and to know when the person needs help, safe pulse-rate zones must first be determined:

- Each person has an individual resting pulse rate, which can be different depending on the person's state of health, fitness level, and natural features. The individual resting pulse rate has to be measured.
- Another step is to determine heart training and a safe pulse rate, which will depend on the person's state of health, fitness level, natural features, and the physical characteristics one wants to train. One training pulse can be useful for the person who wants to train for endurance while another is useful for maintaining the current state of health, and so on.

- The maximum pulse rate should also be determined, which could be reached one or two times per therapy training. This could reach a critical rate if it lasts for longer than 3 to 5 minutes or it is reached more than three times during a therapy session.
- The state of the cardiovascular system can be characterized not only by pulse rate but also by pulse recovery time from raised back to normal again, which is why this data should also be determined.
- The critical pulse point and critical recovery time should always be defined.
- The evaluation of pulse rate using the formula
- [M]pulse rate $= 220 - $ age

  can be used only for completely healthy people. This evaluation is quite simple if other health factors are not kept in mind. That is why more comprehensive cardiovascular and pulmonary system tests for sanatorium clients should be used. For initial individual pulse-rate zone determination, cardiopulmonary exercise testing could be chosen, which includes

    - **Measurement of V'O2max and definition of the anaerobic threshold (AT)**. Objective and individual determination of training pulse rate ranges; creation of health- and performance-oriented exercising schedules
    - **Fat burning test**. Visualization of fat and carbohydrate burning during workload increase; determination of pulse-rate range for maximum fat burning; controlled weight loss through effective metabolic fat exercising
    - **Resting metabolic rate test**. Measuring the metabolic rate at rest for controlled energy uptake; visual comparison of current outcome and targeted value; nutritional recommendations to optimize body weight

- Depending on the test performed, the program offers the relevant results, comments, and recommendations.

This application scenario is very important in the sense that its benefits are tangible to the end user in the following ways:

- Data registration determines safety and better quality of service.
- It provides easier management of complicated treatment services.
- Optimization of the treatment process prevents useless reduplication of procedures.
- It results in more standardized work processes and fewer mistakes.
- The data analysis optimizes the administration processes and resources planning and consumption in the facilities.
- It allows for efficient therapy and workout time.
- It gives an accurate objective of a client's overall health and physical status for a potential evaluation.
- It provides quantitative and qualitative control and evaluation and analysis of the therapy process.
- It eliminates the human factor in evaluation, treatment, and conclusion.
- It provides easy access to all kinds of information.

Table 8.3 Vital Sign and Environmental Monitoring Application's User Requirements

Parameter	Preferred Features
Mobility	Fixed beacon sensor nodes and mobility wearable sensor nodes
Energy resources	Rechargeable mobile nodes, battery-powered fixed nodes and mains
Deployment	Manual placement
Heterogeneity	Body sensors, emplaced sensors, PDAs, data receiver, environmental sensors
Scalability	Medium number of possible future nodes
Topology	Star, mesh, multi-hop, single-hop, hierarchical cluster-based
Coverage	Sparse and dense
Connectivity	Medium and full network connection
Communication modalities	RF-based
Infrastructure	WANs, monitoring nodes, cluster members, cluster heads, gateways, base station, PDAs
Sensor cost	Low-cost *
Sensor type	Tiny, nonintrusive, easy to deploy, easy to wear, robust
Network size	Tens to hundreds of nodes
Network lifetime	Months to years
Application-dependent QoS parameters	Data reliability, time delay (real-time needs), data privacy and security, real-time data processing, fault tolerance, low end-to-end delay, time synchronization

*Sensor nodes cost from 100€ to 400€ each.

- It features universal system adaptability.
- It provides remote control of the system using an Internet connection.
- The information could be used for scientific research.
- A nurse calling system could be installed in clients' rooms and around the building.

Table 8.3 shows the preferred features for the critical vital sign and environmental monitoring application scenario.

8.2.3.3 Technical Analysis

This section identifies and analyzes the use cases and corresponding actors of the critical monitoring application scenario. Based on this information, the technical requirements for the final system are extracted.

Actors

Client: represents the sensor nodes worn by the clients. The sensor node will provide pulse-rate measurements and send an alarm to the client by a beeper. The sensor node may also be a sensor deployed in the environment, responsible for providing temperature and humidity measurements of the environmental conditions in the sanatorium.

Event scheduler: represents the main component that handles actions received as events, automatically schedules them, and coordinates the functional components of the system. This actor is responsible for processing raw data and cross-referencing it to the data already stored, for storing the processed data, and for checking the nodes' status. This actor handles the health alarm rules in terms of the execution and generation of health recommendations and alarms. Finally, it also checks the confirmation of alarm notification.

System administrator: represents the system manager in charge of the system, with additional access rights in order to process clients' personal data or, if he or she is a doctor, with access rights to clients' historical health data. The system administrator may also calibrate the smart tags, register and unregister new clients, manage their information, and recall stored data from the database. Finally, the system administrator is responsible for entering and managing the health alarm rules, which in turn inform the system of the best way to handle the emergency.

Personnel: represents the staff working in the sanatorium, who are in charge of clients' health and treatment. As such, the personnel may request and edit a client report for the presentation of stored data and will confirm alarm notifications upon reception. The personnel in this scenario may also request stored vital sign data and environmental data. He or she can also list the defined health alarm rules for consultation.

Use Cases

Node status checking: The event scheduler may periodically check the status of the wireless nodes to identify failures and notify the system administrator.

Register/unregister new client: For any new client, the system administrator registers the relative information in the system database, forming a corresponding record for later use. For any registered client, the system administrator has the right to unregister its corresponding record and remove it from the system database.

Manage client information: For any registered client, the system administrator may modify the stored information in its corresponding record in the system database. The personnel may also change the information of the active client and store the updated report.

Smart tag calibration: The system administrator may calibrate the smart tag when necessary.

Pulse-rate measurement provisioning: The sensor node worn by the client continuously monitors the pulse rate and sends the measured data with the help of fixed sensors to the central server. For this procedure, an efficient routing algorithm is necessary to guarantee fast and successful data transfer through the network.

Store process data and vital sign raw data: The event scheduler captures all just-processed data of vital signs and environment. The information is stored permanently in the system and remains available for future demands.

Presentation of clients' current, historical, or statistical vital sign data: The personnel may demand information from the system to check clients' current or historical vital sign data. If the client is active, the system checks, extracts, and sends the requested information to the event scheduler. The personnel may also check clients' cross references or statistics. Cross references represent a combination of different types of data during the same interval of time.

Manage health alarm rule: The system administrator is responsible for entering or deleting a health alarm rule for every client, which can set off an alarm. The system administrator may also modify an existing rule by changing the Parameters for activating the alarm. Such Parameters may be the behavior pattern, the action recommended, and the type of alarm.

List health alarm rules: Personnel may see the complete list of health alarm rules as well as their details.

Execute health alarm rules in processed data: The event scheduler makes a request to the system to check whether or not the processed data activate an alarm rule. The system notifies the event scheduler of whether or not an alarm pattern has been matched that will set off an alarm.

Alarm indication: When a rule has been executed and its pattern is matched, the event scheduler makes a request to the system to check the alarm type. The system sends an alarm to the personnel's PDA or PC and a beep to the client's sensor node. Alarms may require immediate attention, early attention, or just consist of a warning. If no confirmation is received within 10 seconds, the system searches for the staff member nearest the client and continues with the notification process until a specialist has been allocated to help the client.

Confirm alarm notification: In abnormal situations, the system is responsible for sending an alarm notification to the personnel's PDA. If the staff member can handle the emergency, he or she sends an affirmative message. Otherwise, if the system receives a negative response or no response, then it notifies the next-closest personnel to the emergency.

Environmental measurement provisioning: Fixed sensor nodes are responsible for sensing the temperature and humidity of the environment and reporting these measurements to the system. The environmental raw data are periodically received from all fixed nodes deployed in the corresponding area.

Stored environmental data presentation: Personnel may request stored environmental data checking. The system extracts, calculates, and sends the requested information either to the personnel's PDA or to the central server.

8.2.4 Technical Requirements

The network topology most suited to support the three scenarios consists of fixed sensor nodes deployed in a grid-based scheme, forming an infrastructure together with a central server that will manage the system database. These requirements are as follows:

- Sensor nodes must communicate with the WSN system.
- Fixed sensor nodes must be part of the WSN system.
- The WSN system must communicate with a PDA.
- The WSN system should communicate with a mobile phone terminal.
- The network should support and, when advantageous, use multiple gateways.
- The WSN may be connected to the WAN through a gateway.
- A main storage system is required.

As far as the networking of the system is concerned, data should be able to be routed securely through the network nodes to multiple destinations within bounded times. When considering the mobile nature of the application scenarios, power conservation is also mandatory as far as network performance is concerned. the details of the communication requirements are as follows:

- The network should support reliable connections.
- The network protocols should be energy-aware and try to minimize energy consumption.
- The network architecture should support authentication and authorization and provide the necessary access control.
- The network architecture may support guaranteed end-to-end latency.
- The system architecture may support QoS guarantees to control delays and provide high transmission probabilities.
- The network architecture should support algorithms and the distribution of cryptographic keys so that the confidentiality of traffic may be protected.
- The networked configuration of sensor nodes and actuators should support reconfiguration as much as possible.
- Network state variables may be available to the application.

With respect to the functionality, the system must provide the following requirements:

- Historical data must always be available and never deleted.
- The system must allow information from surveillance, critical monitoring, and multiple-tracking applications to be cross-checked.
- Only specialized personnel and the system manager shall obtain client reports.
- The historical vital sign data available must be the pulse rate and timestamp.
- Historical vital sign data should be available to specialized personnel.
- Vital sign information must only be available for specialized personnel and the system manager.

- New-client registration time should be less than 15 minutes.
- The following information is required of clients at registration: name, gender, age, resting pulse rate, relevant medical history, and wearable sensor identifier.
- The system manager may access all stored information, register new clients, and unregister clients.
- The system manager and specialized personnel may edit, modify, and consult stored client information.
- The system must detect whether or not the node is activated.
- Smart tags are wearable, and their weight should be less than 100 grams.
- The system should be made up of a network of fixed sensor nodes inside and outside the sanatorium building.
- Smart tags must be weatherproof.
- The smart tag battery must be low-power for a continuous operation of a minimum of 7 days.
- The fixed sensor node must be weatherproof.
- Fixed sensor nodes placed inside the sanatorium building should be connected to the mains.
- Fixed sensor nodes placed outside the sanatorium building must have a continuous operation of a minimum of 30 days.
- The smart tag battery must be rechargeable.
- The system should store all new transmitted sensor node data.
- Node data transmitted must contain a timestamp.
- Node data transmitted must be encrypted.
- Critical data sent should reach the system within 10 seconds.
- Clients should be able to sleep with the sensor node on.
- Smart tags must have a setup and calibration time of less than 15 minutes.
- Data frequency update must be less than 5 seconds.
- Smart tags must have a sensor node to measure pulse rate.
- Smart tags must measure pulse rate continuously when needed.
- Smart tags must report pulse rate precisely every 60 seconds.
- Data collected must be prioritized, with vital sign measurements being the most critical data, while environmental data have a lower priority.
- When an alarm is activated, a client's smart tag must start to beep.
- The alarm beep of the client's smart tag should stop after 5 seconds.
- The alarms should be categorized according to the priority of the action response.
- When an alarm is activated, specialized members of the personnel nearest the client must receive a warning in their PDA or mobile phone with the emergency level of the alarm and the recommended action.
- Recipients must confirm reception of an alarm notification.
- The system must ensure reception of all alarm notifications.
- Raw data must be processed.
- The user interface should display the last update of requested data.
- All processed data must be stored in the system.
- Processed pulse-rate data shall be displayed on a user-friendly interface.

- Processed vital sign data shall be displayed in the PDAs or PCs of specialized personnel.
- Processed vital sign data should be displayed upon request by specialized personnel.

Finally, considering the rest of the characteristics a system must provide, namely reliability, usability, performance, maintenance, and documentation, the following requirements may be extracted from the use case analysis of the three application scenarios.

Usability

- Sensor node should support an alarm signal.
- PDA must support an alarm signal.
- The mobile phone terminal must have an alarm signal.

Reliability

- The WSN system must be available 100% of the time.
- Average time between failures should be greater than 200 hours.
- Average time for repairing the WSN system should be two hours.
- The sensor node's precision should be at least
    Pulse rate: ±0.1% (heart beat per minute)
    Temperature: ±0.5%
    Humidity: ±0.5%

Performance

- The maximum transition response time between the reporting of critical vital sign and the signaling of the alarm must be 1 second.
- The server must be capable of processing a minimum of 20 transmissions per second between the sensor node and the WSN system.
- Changes in the density of WSN nodes shall not affect both the response time and the throughput of the system.
- In degraded mode, the measurement and transmission of vital signs are required.

Maintenance

- Software utility is required for testing the linkage state of the sensor node.
- Software utility should be available for testing the linkage state of the PDA.
- Software utility is required for testing the linkage state of the fixed sensor node.

Documentation

- Installation and configuration guide is required.
- Sensor node user manual is required.
- PDA user manual is recommended.
- WSN system management manual is required.

Additional functional requirements for each specific application scenario are described below, capturing the particular details of tracking accuracy, surveillance criticality, and vital sign monitoring and alarm handling.

8.2.4.1 Multiple-Target Tracking

- The system should allow client position reports to be obtained.
- Historical positional data available must consist of the position, client ID, and timestamp.
- The system should allow client positional statistics to be obtained.
- Positional information must be available only for specialized personnel and the system administrator.
- The tracking algorithm's accuracy must be 5–7 meters.
- The client's maximum speed shall be 10 km/h.
- The maximum number of clients moving outdoors shall be 100.
- The maximum number of clients moving indoors shall be 20.
- The client's position must be tracked with an overall error of 10–15 meters.

8.2.4.2 Surveillance

- The system should allow intruder position reports to be obtained.
- Historical perimeter data available must consist of the perimeter zone, personnel ID, and client ID.
- The system should allow perimeter statistics about the personnel, client, and intruder to be obtained.
- The system must use alert rules to recommend actions and send alarms.
- Only the system administrator may manage security alert rules.
- Security alert rules should consist of recommended actions and associated alert types.
- Only verified rules may activate an alarm.
- Perimeter information should be available only for specialized personnel and the system administrator.
- The perimeter must be continuously observed to detect intruders in real time.
- Fixed sensor nodes in the perimeter of the sanatorium building must constantly control the entrance/exit.

8.2.4.3 Critical Monitoring

- The system should allow client reports to be obtained.
- The system should allow clients' vital sign statistics to be obtained.
- The system should allow statistics comparing clients' vital sign and environmental data to be obtained.
- The system must allow health alarm rules to recommend actions and send alarms to be used.

- The system should allow health alarm rules such as create, edit, modify, list, consult, or delete to be managed.
- Only the system manager may manage health alarm rules.
- Health alarm rules should consist of patterns, actions recommended, and associated alarm types.
- Health alarm rules must check stored data every 5 minutes.
- Only verified rules may activate an alarm.
- The historical environmental data available must be the temperature, humidity, node location, and timestamp.
- Environmental information must be available for all personnel and the system manager.
- Fixed sensor nodes inside the sanatorium building must measure temperature.
- Fixed sensor nodes inside the sanatorium building must measure humidity.
- Fixed sensor nodes inside the sanatorium building must measure temperature precisely every 15 minutes.
- Fixed sensor nodes inside the sanatorium building must measure humidity precisely every 15 minutes.
- Fixed sensor nodes inside the sanatorium building should report temperature precisely every 60 minutes.
- Fixed sensor nodes inside the sanatorium building should report humidity precisely every 60 minutes.
- There must be a configurable alarm based on the allowance of frequency while training.
- Processed humidity and temperature data may be displayed together or separately in a user-friendly interface.
- Processed environmental data may be displayed in personnel's PDAs or PCs.
- Processed environmental data may be displayed upon request of the personnel.

References

Akyildiz IF, Stuntebeck EP (2006) Wireless underground sensor networks: Research challenges. *Ad Hoc Netw* 4(6):669–86.

Anderson E, Girard T, Ottavianelli G (2003) A micro-satellite and in situ ground-sensor network for combating malaria. In *Proceedings of the 54th International Astronautical Congress of the International Astronautical Federation, the International Academy of Astronautics, and the International Institute of Space Law.*

Baumgartner K, Robert S (2006) Architecture of a scalable wireless sensor network for pollution monitoring. In *Proceedings of the 3rd European Workshop on Wireless Sensor Networks.*

Beckwith R, Teibel D, Bowen P (2004) Report from the field: Results from an agricultural wireless sensor network. In *Proceedings of the 29th Annual IEEE International Conference on Local Computer Networks,* pp. 471–478.

Biagioni ES, Bridges KW (2002) The application of remote sensor technology to assist the recovery of rare and endangered species. *Int J High Perform Comp Appl* 16(3).

Bruno MS, Blumberg AF (2006) The Stevens Integrated Maritime Surveillance and Forecast System: Expansion and enhancement. Stevens Institute of Technology. ONR Grant N00014-03-1-0633.

Chang E, Wang YF, Zhao F (Conference chairs) (2004) *Proceedings of the 2nd ACM International Workshop on Video Surveillance & Sensor Networks*, ACM, New York.

Coleri S, Cheung SY, Varaiya P (2004) Sensor networks for monitoring traffic. In *Proceedings of the 42nd Annual Allerton Conference on Commuinication, Control, and Computing*.

Conant R (2006) Wireless sensor networks: Driving the new industrial revolution. *Indus Embed Syst* (Spring/Summer).

Crossbow Technology Inc. (2007) MICA2 868, 916 MHz. http://www.xbow.com/Products/ productsdetails.aspx?sid = 72. Accessed December 2007.

Defense Advanced Research Projects Agency, Joint Program Steering Group Arlington, Virginia, NISE East Electronic Security Systems Engineering Division, North Charleston, South Carolina (1997) Perimeter Security Sensor Technologies Handbook (online manual). http://www.nlectc.org/perimetr/start.htm. Accessed January 2007.

Evans J, Raychaudhuri D, Paul S (2006) Overview of wireless, mobile and sensor networks in GENI. GENI Design Document 06-14, Wireless Working Group.

Gao T, Greenspan D, Welsh M, et al. (2005) Vital signs monitoring and patient tracking over a wireless network. In *Proceedings of the 27th Annual International Conference of the Engineering in Medicine and Biology Society (IEEE-EMBS 2005)*, pp. 102–105.

HEARTS R&D Project. Health Early Alarm Recognition and Telemonitoring System. http:// heartsproject.datamat.it/hearts. Accessed January 2007.

IBM Corporation (2005) Personal Care Connect Mobile Health Monitoring Solution. http:// www-03.ibm.com/industries/healthcare/doc/content/bin/Personal_Care_Connect_12_05_1. pdf. Accessed January 2007.

Kung HY, Hua JS, Chen CT (2006) Drought forecast model and framework using wireless sensor networks. *J Infor Sci Eng* 22(4):751–69.

Lo B, Thiemjarus S, King R, et al. (2005) Body sensor network—A wireless sensor platform for pervasive healthcare monitoring. In *Proceedings of the 3rd International Conference on Pervasive Computing*.

Low KS, Win WN, Er MJ (2005) Wireless sensor networks for industrial environments. In *Proceedings of the International Conference on Computational Intelligence for Modelling, Control and Automation and the International Conference on Intelligent Agents, Web Technologies and Internet Commerce*, pp. 271–276.

Magal Security Systems Ltd. (2002) Barricade 500—Vibration detection system: Advanced outdoor vibration detection system. http://www.magal-ssl.com/products/?pid = 19. Accessed February 2007.

Mainwaring A, Culler D, Polastre J, et al. (2002) Wireless sensor networks for habitat monitoring. In *Proceedings of the 1st ACM International Workshop on Wireless Sensor Networks and Applications (WSNA '02)*, pp. 88–97.

Center for Future Health, University of Rochester (2005) Smart Medical Home Research Laboratory. http://www.futurehealth.rochester.edu/smart_home/. Accessed January 2007.

MobiHealth B.V. (2007) MobiHealth Project. http://www.mobihealth.org. Accessed December 2007.

Pakzad SN, Kim S, Fenves GL, et al. (2005) Multi-purpose wireless accelerometers for civil infrastructure monitoring. In *Proceedings of the 5th International Workshop on Structural Health Monitoring (IWSHM 2005)*.

Pilone D (2006) *UML 2.0 Pocket Reference*, O'Reilly Media, Inc., Sebastopol, CA.

Roark RC, Van Wie DG (2003) A new ALERT protocol: Feasibility study of a new air interface and physical layer packet definition for the ALERT user community. Blue Water Design, LLC.

Sanders JM (2000) Sensing the subtleties of everyday life. *Res Horiz* 17(2).

Sazonov E, Janoyan K, Jhac R (2004) Wireless intelligent sensor network for autonomous structural health monitoring. *Proc SPIE* 5384:305–14.

Sensicast, Inc. (2008) SENSICAST. http://www.sensicast.com/. Accessed January 2008.

Shea DA, Lister SA (2003) The BioWatch Program: Detection of Bioterrorism. Congressional Research Service, Library of Congress, Washington, DC. Report No. RL32152.

Van Laerhoven K, Lo BPL, Ng JWP, et al. (2004) *Proceedings of the 3rd International Workshop on Ubiquitous Computing for Pervasive Healthcare Applications (UbiHealth 2004)*.

Zhao F, Guibas L (2004) *Wireless Sensor Networks: An Information Processing Approach.* Morgan Kaufmann, San Francisco.

Index

Printed in the United States